# On Willing Selves

*Selected books by Sabine Maasen*

GENEALOGIE DER UNMORAL – ZUR THERAPEUTISCHEN KONSTRUKTION SEXUELLER SUBJEKTE (GENEALOGY OF THE IMMORAL – THERAPEUTIC CONSTRUCTIONS OF SEXUAL SELVES) (1998)

WISSENSSOZIOLOGIE – EINE EINFÜHRUNG (SOCIOLOGY OF KNOWLEDGE: AN INTRODUCTION) (1999)

METAPHORS AND THE DYNAMICS OF KNOWLEDGE (with Peter Weingart) (2000)

SCIENCE STUDIES: PROBING THE DYNAMICS OF SCIENTIFIC KNOWLEDGE (edited with Matthias Winterhager) (2001)

VOLUNTARY ACTION: MINDS, BRAINS, AND SOCIALITY (with Wolfgang Prinz and Gerhard Roth) (2003)

DEMOCRATIZATION OF EXPERTISE? EXPLORING NOVEL FORMS OF SCIENTIFIC ADVICE IN POLITICAL DECISION-MAKING (edited with Peter Weingart) (2006)

BILDER ALS DISKURSE? BILDDISKURSE (IMAGES AS DISCOURSES / DISCOURSES ABOUT IMAGES) (edited with Torsten Mayerhauser and Cornelia Renggli) (2006)

# On Willing Selves

## Neoliberal Politics *vis-à-vis* the Neuroscientific Challenge

Edited by

Sabine Maasen and Barbara Sutter
*University of Basel*

First published 2007 by
PALGRAVE MACMILLAN
Houndmills, Basingstoke, Hampshire RG21 6XS and
175 Fifth Avenue, New York, N.Y. 10010
Companies and representatives throughout the world

PALGRAVE MACMILLAN is the global academic imprint of the Palgrave
Macmillan division of St. Martin's Press, LLC and of Palgrave Macmillan Ltd.
Macmillan® is a registered trademark in the United States, United Kingdom
and other countries. Palgrave is a registered trademark in the European
Union and other countries.

ISBN-13: 978-0-230-01343-8    hardback
ISBN-10: 0-230-01343-0    hardback

This book is printed on paper suitable for recycling and made from fully
managed and sustained forest sources. Logging, pulping and manufacturing
processes are expected to conform to the environmental regulations of the
country of origin.

A catalogue record for this book is available from the British Library.

Library of Congress Cataloging-in-Publication Data

On willing selves : neoliberal politics vis–á–vis the neuroscientific
    challenge / edited by Sabine Maasen & Barbara Sutter.
        p. cm.
    Includes index.
    ISBN  0–230–01343–0 (alk. paper)
        1. Will  2. Neurosciences.  3. Self.  4. Political science—Philosophy. I. Maasen,
Sabine, 1960– II. Sutter, Barbara.

BF611.O5 2007
153.8—dc22                                                                2007023344

10  9  8  7  6  5  4  3  2  1
16  15  14  13  12  11  10  09  08  07

Printed and bound in Great Britain by
Antony Rowe Ltd, Chippenham and Eastbourne

# Contents

# Acknowledgements

The book is based on a conference on 'Willing and Doing,' held at the Max Planck Institute for Psychological Research in Munich from 18 to 19 February 2004. It took place as part of an interdisciplinary research program 'On the Nature and Culture of Volition'. We gratefully acknowledge the financial support of our project as well as the conference by the VolkswagenStiftung, Hannover. We owe special thanks to Wolfgang Prinz and the entire research group who supported our endeavor with unceasing interest and friendly provocations. Notably Stefanie Duttweiler enriched our discussion: government of selves and others by pursuing happiness (in science and elsewhere) has become a running theme. Last but not least, we thank Thomas Oehler for diligently handling the manuscript as well as Daniel Bunyard, Palgrave Macmillan, for accompanying the process with competence and care.

# Notes on Contributors

**John Clarke** is Professor of Social Policy at The Open University, UK. His research interests are politics and ideologies of welfare, political, cultural, and organizational changes associated with managerialism in social welfare, impact of consumerism on public services, and the contribution of cultural analysis to the study of social policy. Pertinent publications are *The Managerial State: Power, Politics and Ideology in the Remaking of Social Welfare* (with J. Newman, 1997), *Changing Welfare, Changing States: New Directions in Social Policy* (2004) and *Creating Citizen-Consumers: Changing Publics and Changing Public Services* (with J. Newman, N. Smith, E. Vidler and L. Westmarland, 2007).

**Barbara Cruikshank** is Associate Professor at the Department of Political Science of the University of Massachusetts, USA. Her work focuses on political theory, democratic and feminist theory, technologies of citizenship, Foucault, and governmentality. Selected publications are *The Will to Empower: Democratic Citizens and Other Subjects* (1999) and 'Revolutions Within: Self-Government and Self-Esteem', in David A. Irwin (ed.), *Foucault and Politics* (1993).

**Stefanie Duttweiler** is Research Assistant at the University of Basel, Switzerland. Her research interests are technologies of the self, sociology of counseling and sociology of religion. Publications include *Sein Glück machen. Arbeit am Glück als neoliberale Regierungstechnologie* (*To Make One's Fortune: The Work on Happiness as a Technique of Neoliberal Governmentality*, 2007), 'Ein völlig neuer Mensch werden – Aktuelle Körpertechnologien als Medien der Subjektivierung' ('Becoming a Totally New Person: Recent Technologies of the Body as Media of Subjectivation'), in K. Brunner et al. (eds), *Verkörperte Differenzen* (2004) and 'Beratung' ('Counselling'), in U. Bröckling et al. (eds), *Glossar der Gegenwart* (2004).

**Kenneth J. Gergen** is Mustin Professor of Psychology at Swarthmore College (USA) and a founding member and Board President of the Taos Institute. Gergen also serves as an Affiliate Professor at Tilburg University in the Netherlands, and an Honorary Professor at the University of Buenos Aires. His work focuses on societal practice, dialogical change, relational theory and the self. Pertinent publications are *Relational Responsibility* (with S. McNamee, 1999), *Social Construction in Context* (2001), and *Therapeutic Realities, Collaboration, Oppression and Relational Flow* (2006).

**Alois Hahn** is Professor of Sociology at Trier University, Germany. His research interests focus on religion and culture, religious roots of the civilization process, identity and modernity, history of sociology, health and illness. Since 1995 he has been Professeur associé at the Centre des Mouvements Sociaux (CEMS) of the Ecole des Hautes Etudes in Paris, currently he is also a fellow at the Wissenschaftskolleg zu Berlin/Institute for Advanced Studies. His publications include *Identität und Moderne* (*Identity and Modernity*, edited with H. Willems, 1999) and *Konstruktionen der Selbst, der Welt und der Geschichte. Aufsätze zur Kultursoziologie* (*Constructions of the Self, of the World and of History. Essays on the Sociology of Culture*, 2000).

**Sabine Maasen** is Professor for Science Studies at the University of Basel, Switzerland. Recent research interests include the social construction of (sexual) selves, the dynamics of knowledge, particularly as regards the notion of consciousness, as well as the dynamics of knowledge between different scientific and nonscientific discourses. Pertinent publications are *Genealogie der Unmoral. Zur Therapeutisierung sexueller Selbste* (*Genealogy of the Immoral – Therapeutic Constructions of Sexual Selves*, 1998), *Metaphors and the Dynamics of Knowledge* (with P. Weingart, 2000), *Voluntary Action: Brains, Minds, and Sociality* (edited with W. Prinz and G. Roth, 2003).

**Armin Nassehi** is Professor of Sociology at the University of Munich, Germany. His research interests concern sociology of culture, political sociology, sociology of religion as well as sociology of knowledge and science. Publications include *Differenzierungsfolgen. Beiträge zur Soziologie der Moderne* (*Consequences of Differentiation: On the Sociology of Modernity*, 1999), *Paradox of Risk Control* (edited with T. Hijikata, in Japanese, 2002), *Geschlossenheit und Offenheit. Studien zur Theorie der modernen Gesellschaft* (*Closure and Openness: On the Theory of Modern Society*, 2003), *Bourdieu und Luhmann. Ein Theorienvergleich* (*Bourdieu and Luhmann: A Comparison of Theories*, edited with G. Nollmann, 2004), and *Der soziologische Diskurs der Moderne* (*The Sociological Discourse of Modernity*, 2006).

**Janet Newman** is Professor of Social Policy at The Open University, UK. Her work focuses on analyses of governance, policy, and politics in trying to understand new political formations such as the managerial reforms of the 1990s and politics and policies of New Labour, the modernization of European welfare states, discourses and practices of publics and publicness. Selected publications are *The Managerial State: Power, Politics and Ideology in the Remaking of Social Welfare* (with J. Clarke, 1997), *Modernising Governance: New Labour, Policy and Society* (2001), *Remaking Governance: Peoples, Politics and the Public Sphere* (edited in 2005), *Power, Participation and Political Renewal* (with M. Barnes and H. Sullivan, 2007), *Creating Citizen-Consumers: Changing Publics and Changing Public Services* (with J. Clarke, N. Smith, E. Vidler and L. Westmarland, 2007).

**Nikolas Rose** is Professor of Sociology at the London School of Economics and Political Science, UK. He is Director of the LSE's BIOS Research Centre for the Study of Bioscience, Biomedicine, Biotechnology and Society. Currently his research concerns biological and genetic psychiatry and behavioral neuroscience, and its social, ethical, cultural, and legal implications. Publications include *Governing the Soul: The Shaping of the Private Self* (1989), *Inventing Our Selves: Psychology, Power, and Personhood* (1996), *Powers of Freedom: Reframing Political Thought* (1999) and *The Politics of Life Itself: Biomedicine, Power, and Subjectivity in the Twenty-First Century* (2006).

**Marén Schorch** is Research Assistant at the Department of Sociology at the University of Trier, Germany. Her research interests focus on identity and alterity, adolescence, citizenship/nationality and identification as well as ethnic/autochthonic minorities. She has published *Meine Zukunft bin ich! Alltag und Lebensplanung Jugendlicher* (*I am my Future! Youth's Everyday Life and Life Planning*, with W. Vogelgesang, I. Eisenbürger and A. Haubrichs, 2001) and 'Tests und andere Identifikationsverfahren als Exklusionsfaktoren' ('Tests and Other Procedures of Identification as Factors of Exclusion', with A. Hahn), in *Grenzen, Differenzen, Übergänge: Spannungsfelder inter- und transkultureller Kommunikation* (forthcoming).

**Barbara Sutter** is Research Assistant at the University of Basel, Switzerland. Her work focuses on the social construction of willing selves by practices of citizenship and civil society. Further interests include issues of critical theory and sociology of science. Publications are 'Governing by Civil Society: Citizenship within a New Social Contract', in J. Angermüller et al. (eds), *Reflexive Representations: Politics, Hegemony, and Discourse in Global Capitalism* (2004) and 'Von Laien und guten Bürgern: Partizipation als politische Technologie' ('On Laypersons and Good Citizens: Participation as Political Technology'), in A. Bogner and H. Torgersen (eds), *Wozu Experten? Ambivalenzen der Beziehung von Wissenschaft und Politik* (2005).

**Mariana Valverde** is Professor of Criminology at the University of Toronto, Canada. Her research interests focus on social and legal theory, historical-sociological studies of moral regulation, and the sociology of law. Publications include *Law's Dream of a Common Knowledge* (2003), *Nietzsche and Legal Theory: Half-Written Laws* (edited with P. Goodrich, 2005), *Law and Order: Signs, Meanings, Myths* (2006), and *The New Police Science: The Police Power in Domestic and International Governance* (edited with M. Dubber, 2006).

**Tillmann Vierkant** is Lecturer in Philosophy of Mind at the University of Edinburgh, UK. His research interests are the relationship between narrative and implicit cognitive processing, theories of volition and freedom of the will informed by contemporary cognitive science, and the importance of phenomenal consciousness for practical philosophy. Selected publications are *Is the Self Real? An Investigation into the Philosophical Concept of 'Self' between Cognitive Science and Social Construction* (2003) and 'Between Epiphenomenalism and Rationality', in S. Maasen et al. (eds.), *Voluntary Action: Brains, Minds, and Sociality* (2003).

**Louise Westmarland** is Lecturer in Criminology at The Open University, UK. Her research interests focus on gender and the police, corruption and professional ethics in the criminal justice system, danger, fear, and ethics, as well as on ethnographic research methods in criminology. Publications include *Gender and Policing: Sex, Power and Police Culture* (2001), 'Policing Integrity: Britain's Thin Blue Line', in C. B. Klockars, M. Haberfeld and S. Kutjnak Ivkovich (eds), *The Contours of Police Integrity* (2003), and *Creating Citizen-Consumers: Changing Publics and Changing Public Services* (with J. Clarke, J. Newman, N. Smith and E. Vidler, 2007).

# Introduction: Reviving a Sociology of Willing Selves

*Sabine Maasen and Barbara Sutter*

## The neuroscientific challenge

Most recently, the neurosciences have become fashionable in society. They stimulate all kinds of initiatives: neuro-didactics, neuro-theology, neuro-economy, and neuro-technology inspire educational, spiritual, industrial, or 'neuro-ceutical' approaches to governing selves and society alike. Call it hype or horror: the recent fad, provocatively stated, is ultimately about brainy selves happily living in neuro-society explained by the neurosciences and largely regulated by neuro-technologies. In truth, while this abbreviated storyline is, as yet, only part of the daydreams of single entrepreneurs in the domain of consulting (see Lynch, 2004), it has already affected general media discourse. Various applications of the neurosciences have attracted journalists' interest. More often than not, the findings have been glossed over in a more humoristic note. Neuro-marketing, for instance, is being mocked as revealing insights that are not really new with the help of costly experimentation and instrumentation. Above all, neuro-marketing seems to have found out that favorite brands are favored by our brains, too (Schnabel, 2003).

The debate becomes far more serious, however, once other applications are being discussed: should we conceive of criminals as responsible and guilty or 'just' as deviant and dangerous? Should we educate our children according to their self-proclaimed needs and talents as their developing brains always know best what they need (e.g., Singer, 2002; Roth, 2003)? Issues such as these are not only treated as more or less probable *applications* of brain research. Rather, they are debated as highly questionable, albeit as yet only thinkable, *implications* of brain research for notions of person, responsibility, and, last but not least, free will. The latter implication, in particular, proves disquieting for most.

1

Thus the self, the conscious, volitional individual and its agency have come under attack – and not for the first time! About twenty years ago, postmodernists had already declared the death of the subject. In this view, the self is a fiction invented *post hoc*. Rather than acting on the basis of 'own' intentions, motives, or habits, the subject came to be thought of as driven by relations of power and knowledge. To postmodernists, these very intentions, motives, or habits are not the sources of a pure and self-conscious individuality, but the result of subconscious desires and discourses external to the self. The subject is merely a crossing of these structures.

Although this debate has not been resolved, other issues such as agency and empowerment (or lack thereof) have become more prevalent concerns. These debates rest on the assumption that the self, the acting, willing, and reflecting self, needs to be more than just a crossing of desires and discourses. At least, the self is said to have, and indeed cultivate, a capacity to reflect and sort out inner and outer conditions for his or her actions, rather than being fully determined by them. To this end, all kinds of educational, therapeutic, and self-instructing practices emerged, currently flooding Western societies: this recent hype of self-help indicates the creed that the self can be advised to determine his or her own actions – within the limits of inner or outer constraints, of course.

To be sure, this call for self-regulating, willing selves is a political one. This is true in two respects: on the one hand, juridical, political, and moral practices, ascribing guilt, enticing active citizenship, or attributing responsibilities in everyday life can hardly be understood without recourse to a subject who consciously steers his or her own courses of action. On the other hand, selves capable of steering themselves and others are said to be not only a condition but also an effect of a society that is increasingly regarded as neoliberal. The link between the two complementary aspects, in our perspective, is to be found in technologies that contribute to a managerial regime we may view as the utmost signature of contemporary selves and society alike. This managerial regime rests on the individuals' capacity for managing themselves and others: their individual happiness, their families, their jobs, their civic engagement, and so on. Ultimately, this capacity capitalizes on a resource that seemed to be long forgotten: the will.

The academic reaction is mixed. While most representatives from disciplines such as theology, philosophy, or pedagogy largely insist on the existence of a free will, the neuro-cognitive faction counters our long-cherished intuitions about what volition and consciousness are, thus challenging this 'inner feel', this special quality accompanying perception, volition, reflection, and action. Inspired by the 'cognitive turn' in psychology, the 'decade of the brain', and the institutionalization of the 'cognitive sciences', neuro researchers insist on consciousness and the notion of free will (or the representation thereof) as being nothing but the name for the interplay of various regions in the brain on which cognitive functions operate. Willing and doing are fully determined by the brain. Other than that, talking about experiential or phenomenal aspects of consciousness and free will is relegated to folk psychological wisdom. Now that heuristics, techniques, and instruments are available, the 'issues proper', that is, the neural correlates of consciousness and free will, can and should be addressed (Milner, 1995; Crick and Koch, 1998; for an overview see, e.g., Zeman, 2001).

By contrast, the social sciences have, thus far, only rarely commented on this debate, neither publicly nor academically. This is all the more astounding, as sociology, from its inception, has been concerned with the issue of voluntary action. Therefore, the introduction will first outline a historical overview of what might be called 'a sociology of voluntary acts' (Section 2). Pursuing such a line of thought, as governmentality studies do, in our view, is paramount to enrich the debate about the neuroscientific challenge – a challenge which amounts to not only deliberating and assessing the role of a (future) key technology, as has just been done, for instance, in a major European technology assessment (Meeting of Minds, 2005) (Section 3). What is more: as it is accompanied with promises and risks alike, the debate should also consider the more fundamental concepts of agency, autonomy, responsibility, hence the notion of willing selves (Section 4). For, governmentality analysis is particularly targeted at our notions of enterprising selves as key to modern governance structures of neoliberal society. It is thus governmental analysis, rather then sheer technology assessment, which is called for (Section 5). The contributors to this book, while not all inspired by a governmental framework, still subscribe to the currently changing role of selves who are assigned with ever more responsibilities concerning the government of themselves and others (Section 6). The final section briefly discusses the role of the neurosciences both in the developments as well as for this book: it is a starting point, an exemplar, yet not the subject matter of this book.

## Voluntary acts as conceived by sociology

From its inception, sociology has assigned voluntary actions a basic role by framing them within the relation of individual action and social structure.

This relation has been studied for variables such as social class, ethnicity, and gender, as well as with respect to topics such as socialization processes, interaction contexts, and culture. In general, the approaches adopt mediating positions between social determinism and methodological individualism, balancing the relative strength of social structure and individual volition.

Most classical sociologists emphasized the importance of structure over agency. Karl Marx (1818–1883), for example, stressed man's alienation from work and society, causing his alienation from himself. On a different note, Émile Durkheim (1858–1917) argued for the centrality of social facts over individual volition: sentiments, moralities, and behaviors could be explained as social facts linked to objective features such as social organization, societal differentiation, and social change. Georg Simmel (1858–1918), however, highlighted both the connections and tensions between the individual and society: the socialized individual is incorporated within society and yet stands against it.

Symbolic interactionism held that a person's self grows out of a person's commerce with others. Voluntary acts, according to Charles Cooley (1864–1929), arise in social processes of communicative interchange reflected in a person's consciousness. From these processes, following George Herbert Mead (1863–1931), emerge voluntarily acting selves composed of three elements: a 'me', who consists of those attitudes of others incorporated into the self, and an 'I', who organizes the attitudes of others, selects objects on which the individual will act, and chooses or commits itself to respond in a certain way. A third element is the 'generalized other', incorporating overarching group values into the individual's appreciation of self.

Phenomenologists posit that typified action and interaction become 'habitualized'. As meaning-striving beings, humans create theoretical explanations and moral justifications to legitimate their conduct. When internalized by succeeding generations, the conduct is institutionalized and exerts compelling constraints over individual volition.

In more recent times, (post)structuralist and postmodernist accounts have prevailed. According to Michel Foucault (1926–1984), volition is based on discourses embedded in the activities, social relations, and expertise of specific communities, whether these are scientific, political, or virtual. Voluntary acts are inseparable from action, environment, access, and empowerment. Pierre Bourdieu (1930–2002) used the term 'habitus' to describe a set of schemes along which individuals conceive their reality and act according to their position within social structure. In theories of structure and agency, such as that of Anthony Giddens's theory of structuration, both variables influence each other equally. In postmodern understanding, however, as advanced by Jean-Francois Lyotard (1924–1998), voluntary acts are, for the most part, symbolically determined: they emerge from culturally available narrative forms, discourses, practices, representations, stories, and images.

Today, so-called governmentality studies restate the role of structure and agency anew: in power decentered neoliberal societies, its members play an increasingly active role in their own self-governance. Governmentality studies are thus preoccupied with the analysis of different societal domains (politics, economy, science, and so on) that, while having their own logics, are highly dependent on voluntarily acting selves. Volition, here, emphasizes both *commitment* to stick to decisions one has made and *change*, if internal and external conditions so require.

The recent call for individual volition is political, and does not challenge the basic sociological outlook according to which voluntary acts are structured by socio-historical conditions and symbolic forms related to them. It is in this spirit that this volume aims at reviving the sociological interest in individual volition – as a building block for contemporary attempts at rearranging the link of selves and society by way of creating neosocial selves to which advances in modern sciences and technologies contribute their share. Thus far, however, the neuroscientific challenge has, if at all, been discussed as one of several key technologies, rather than as an indicator, if not vehicle to change our notions of selves and society. While the attention drawn by the neurosciences ensures the necessary publicity of the topic, this focus also narrows the scope of the debate.

## Deliberating a key technology

'If we are entering a new age for public understanding of science', Miller writes (2001, p. 119), 'it is important that citizens get used to scientists arguing out controversial facts, theories, and issues. More of what currently goes on backstage in the scientific community has to become more visible if people are going to get a clearer idea of the potential and limitations of the new wonders science is proclaiming' (see also Hilgartner, 2000). Advances in science and technology change our interaction with one another and thereby affect 'how we define the general welfare, our views on the role of government, and the approach we take in exercising our rights as global citizens' (Fowler, 1999, p. 153). At present, scientific and technological progress allows for interventions in spheres we could not attain before. These opportunities confront us with the necessity of taking decisions.

Particularly, choices emerging from advances in the knowledge about organisms (non-human and human) face us with a multitude of ethical, legal, and social decisions to be taken. Among the issues currently regarded as most pressing are: privacy, handling of prenatal life, manipulation of germ lines, free will and individual responsibility, genetic reductionism and determinism, and genetic enhancement. 'Fundamental to all is the realization that there is no perfect or ideal human genetic constitution to which all others are inferior, as well as the recognition that genetic information should not be used to advance ideological or political agendas' (Fowler, 1999, p. 153). In order to assess risks and chances accompanying novel scientific and technological developments, a knowledgeable public is considered a prime requisite.

In this vein, the recent neuroscientific challenge of free will can be read as an instance of public understanding of science. Particularly in Germany, but also elsewhere, neuroscientists eagerly enlighten the general public about their field (Meeting of Minds, 2005). The common narrative is that the complexity of the brain is so great that understanding even the most basic function will require even more research – and more funds. Notably, neuroscientists alert us to the therapeutic promises involved in their research. For instance, by understanding how neurons grow, one could develop new technologies for central nervous system repair following neurological disorders and injuries.

However, similar to debates on other key technologies (such as gene technology or stem cell research), for the neurosciences, too, public acceptance is hard to attain. Even more so, as promises and perceived threats strongly interact with pre-existing knowledge (false or correct). Above all, it connects to the fear of 'being determined,' this time by neurophysiological processes: it deals with technological promises that are met with ambivalences (e.g., neurodidactics), and, most importantly, it bears on the critical concepts of responsibility and free will underlying social interaction, in general, and legal decisions, in particular. The neuroscientific findings thus challenge the understanding of persons as both moral selves and politically acting citizens. This, however, is only rarely discussed. If such debates do occur, they largely lack both publicity and the participation of the social sciences (Maasen, 2006b). All the more reason to start filling the gap.

## Willing selves

Ironically enough, just at the time the neurosciences declare free will an illusion, societal practices set out to call for autonomously choosing actors.

While this coincidence will be discussed at the end of this introduction, it is worthwhile to look more closely now at the notion of free will, autonomy, and agency. In what way may the analysis of willing selves contribute to understanding the current condition of selves and society?

Underlying neoliberal governmentality is a more general vision that every human being is an entrepreneur managing his or her own life and should act accordingly. This mode of subjectivation can occur in various ways (directly and indirectly), but one of the main modalities is 'interpellation' as described by Louis Althusser. It is by accepting entrepreneurship as a task each individual should assume that the (excessively) entrepreneurial selves come about. Again, this acceptance is not an individual idiosyncrasy but rather a social affordance – a whole network of practices, routines, institutions, organizations covering virtually all domains of life (e.g., education, workplace, legal system) both ask for and co-produce the entrepreneurial self. It is about to become a standard view in society about how the desirable subject has to be: neoliberal society favors a subject that calculates rationally and acts responsibly.

No wonder, then, that the concept of the entrepreneurial self is to be found at the bottom of all ideas of neoliberal subjectivation. In terms of moral philosophy, as Patrick Fitzsimons points out, 'this amounts to a "virtue ethic," in which human beings are supposed to act in a particular way according to the ideal of the entrepreneur. Individuals who choose their friends, hobbies, sports, and partners to maximize their status with future employers are ethically neoliberal' (Fitzsimons, 2002, p. 3). For many observers of neoliberal society, this represents a spread of the market principle into virtually all domains of life: the market or market-like structures are seen as the guiding principle of non-economic spheres as well. However, Fitzsimons, more pointedly than other observers of neoliberal society, notes that for 'neo-liberals it is not sufficient that there is a market: there must be nothing which is not a market' (2002, p. 3).

This market requires the autonomous chooser, a self, that is capable of acting in both a responsive and responsible way toward the ever-unruly environment – the buzz words being: autonomization of society, deregulation, privatization, ubiquitous and continuous influx of information by, for example, new technologies, the media, and public debate. All of this permeates personal life, demanding perpetual and flexible responses – *voilà* enterprise culture. Originally stemming from the domain of economic liberalism, it has become a *leitmotif* for current culture as a whole, now 'concerned with the attitudes, values, and forms of self-understanding embedded in both individual and institutional activities' (Keat and Abercrombie, 1991, p. 1). Governmental practices, guided by institutions and programmed by organizations, typically rely on the assumptions of methodological individualism. The enterprise culture puts the individual in a continuous exercise of self-reformation to be what it must become in order to deal with what the next technology demands. The emphasis is on the individual's competence to evaluate the environment as well as its effects on it (see Fitzsimons, 2002, p.7).

## Analyzing neoliberal governmentality

As 'it's true that "practices" don't exist without a certain regime of rationality' (Foucault, 1991, p. 79) analyzing governmental practices, *pace* Foucault, implies exploring the style of thought with which they coincide. Questions such as 'who can govern? what is governing? what or who is governed?' allow for an insight into the conditions of, and constraints on, the capability of rendering governmental activities 'thinkable and practicable both to its practitioners and to those upon whom it was practised' (Gordon, 1991, p. 3). In this vein, Foucault extends the analysis of government: apart from the exercise of political sovereignty, it concerns 'the relation between self and self, private interpersonal relations involving some form of control and guidance, relations within social institutions and communities' (Gordon, 1991, pp. 2–3).

The target of this analysis is not institutions, theories, or ideology, but governmental practices,

> the hypothesis being that these types of practice are not just governed by institutions, prescribed by ideologies, guided by pragmatic circumstances – whatever role these elements may actually play – but possess up to a point their own specific regularities, logic, strategy, self-evidence and 'reason'. It is a question of analyzing a 'regime of practices' – practices being understood here as places where what is said and what is done, rules imposed and reasons given, the planned and the taken for granted meet and interconnect.
>
> (Foucault, 1991, p. 75)

In particular, technologies of domination and technologies of self are, thus Foucault, the prime techniques used 'to make the individual a significant element for the state' (Foucault, 1982, p. 153). He asks which technology of government has now been put to work in order to make the individual a significant element other than the earlier technique termed 'policing' (Foucault, 1988, p. 153). Which strategy, self-evidence, or logic rendered it acceptable for the individual to be a (self-) governed element for the state? From this perspective, all practices concerned with (self-) education, (self-) management, (self-) therapy, or counseling can be regarded as pivotal governmental techniques in that they perfectly coalesce technologies of domination (e.g., the wish to educate) and technologies of self (e.g., the wish to become educated). In a controlling and, at the same time, disciplining manner, these technologies present the relay between the (self-) reforming self and the reforming social institutions they live in or work at. Selves and society co-produce their market-like transformations that ultimately are about creating enterprising selves in an enterprise culture.

This calls for deciphering the discursive regime, that is, 'the processes, procedures, and apparatuses whereby truth, knowledge, and belief are produced' (Fraser quoted in Fitzsimons, 2002, p. 6), for it is through those 'productions of truth, knowledge, and belief ... that we become both "governed and governable"' (Fitzsimons, 2002, p. 5). It is the working of these apparatuses which become the object of govenmentality studies, as they are seen as being the building blocks of neoliberal politics.

In neoliberal society the emphasis is, hence, on the individual's autonomy: in a continuously changing environment, individuals are invited to become a 'subject of change' rather than being or feeling 'subjected to change'. Although this can easily be dismissed as 'rhetoric' and mere 'ideological superstructure', a Foucauldian perspective, as outlined above, makes us see rhetorics as embedded in institutional as well as organizational settings, routines, and procedures that bring the enterprising self into existence – voluntarily, that is, albeit, more often than not, *nolens volens*. Ongoing improvement of one's skills and capabilities need not be enforced by sheer force, rather it is market-like demands such as employability and marketability leading the individual to comply – voluntarily. To the extent that individuals adhere to this new culture of being and new culture of work, enterprising selves as well as enterprising culture become real.

*Peu-à-peu*, all non-economic domains of life are subjected to this voluntary mode of subjecting oneself and others to market-like structures and demands. An 'economization of the social' has been diagnosed (Bröckling et al., 2000). Education, work, legal system, and everyday life: though different in pace, all these more or less significant shifts contribute to nothing less than a new cultural matrix: neoliberal governmentality. The exercise of power relations in the emerging governmental mode firmly establishes a relation between self and itself, interpersonal as well as relations within social institutions and community that are ultimately based on control – a control based on governing oneself and others in such a way that technologies of domination and technologies of self coalesce in concepts of autonomous choosing or voluntary acting.

## On the making of willing selves

The contributions assembled in this volume all subscribe to the notion that the self and the technologies of making, shaping, if not inventing it, are an accomplishment both by and for our society. While not all contributions are directly inspired by Foucault's work on governmentality and though some authors are skeptical about accordant approaches, they nonetheless share the basic insight that the astounding acceptability of self-technologies in ever more domains of government is a joint venture achieved at various sites in society. In the heterogeneity of issues and avenues explored, they also subscribe to the basic assumption that the relation between theories and practices of government need careful differentiation. It is necessary to analyze and discuss the interconnection between domains of government and self-formation (Dean, 1994, p. 159). They differ markedly across domains, and with respect to the prime sources of governmental activities (initiated by authorities or the individuals themselves). As to the latter, one has to acknowledge the fact that dominant discourses might meet with alternative identifications of individuals (Clarke, Newman and Westmarland, Chapter 5 in this volume).

This requires a variety of approaches, disciplinary perspectives, issues, and methods. The contributions in this volume represent this variety, albeit in a limited, yet, open-minded way, always considering neighboring perspectives. In accordance with the structure of the overall argument, the book is divided into four parts, each predominantly probing into one particular aspect: the cultural evolution of self-technologies, the role of (novel) (social-)scientific knowledge about selves, the (neo-)politics of self, the morality and ethics involved in the notion of (self-governing) selves. All four parts are headed by a brief introduction tying them into the overall argument of the book: put simply, neoliberal societies are characterized by the creation of self-regulating selves able to govern themselves and others. It is by way of various technologies (e.g., of agency, of performance, of citizenship) that these selves thus become not only highly individualistic but highly sociable as well: we are observing the making of neosocial selves. Ultimately, all technologies rest on what we call technologies of the will, now increasingly incorporated into the managerial regimes of neoliberal society.

As most prominent forms of conduct, neoliberal governance presupposes the individuals' capacity to determine, regulate and govern themselves, thereby fusing ethical and political domains (Dean, 1995, p. 562). 'Governmental-ethical practices,' in the words of Rimke, 'underline the way in which what might loosely be called "practices of government" come to depend upon, and operate through, "practices of the self", such as those in popular self-help texts' (Rimke, 2000, p. 71).

This is precisely where the narrative of this volume begins: in Chapter 1 'Self-Help: The Making of Neosocial Selves in Neoliberal Society', Sabine Maasen, Barbara Sutter and Stefanie Duttweiler start out by observing that in our (post-) modern societies the willing self has come to the fore, neuroscientific denouncements notwithstanding. A brief genealogical study of self-help manuals reveals them to be part and parcel of those more or less rationalized programs and policies teaching us to govern ourselves and others – not just by way of control or discipline, but also in order to make ourselves or others happy, virtuous, and so on. In neoliberal regimes, the government of selves is best linked, so it seems, to practices in which free individuals are enjoined as simultaneously subjects of liberty and of responsibility. Interestingly enough, self-help literature, while (implicitly) referring to psychological knowledge, is not to be conceived as a deviant form of expert therapeutics, but rather to be regarded as a genre *sui generis*. This shows in the way this literature branches off into ever more special issues that interlink with specific demands in neoliberal society. While one, seemingly more self-centered type, differentiates into virtually all domains of everyday life (health, happiness, youth, success), a more recent branch of handbooks provides 'education for citizenship' in order to attain the 'communally responsible person.' On the other hand, the self-help literature draws from various sources, such as manuals educating our manners as well as from practices of counseling and psychotherapy, to mention but a few. Yet, the respective authors all take care to reframe these practices as self-technologies such that they lend themselves to guiding the self in an environment that is irredeemably turbulent and in constant need of adaptive responses by the individual.

To accomplish this task, modern self-helpers need to exercise honesty with themselves. Honesty is regarded as instrumental for initiating real self-transformation and is embedded in techniques based on self-monitoring. Hence, the requirement to confess the truth about one's self is typically presented as a practice integral to self-diagnosis. Confessional practices represent a powerful cultural conviction of the ability to both liberate and regulate oneself, with or without the help of experts. In order to understand the self-evidence and cultural acceptability of such practices, it is most enlightening to trace their genealogy, as Alois Hahn and Marén Schorch do in Chapter 2, 'Technologies of the Will and Their Christian Roots'.

The quasi-juridical regime of confession licensed two crucial developments in Western culture: a new Christian technology of the self and a discourse tailored to the requirements of power of the medieval church. The Christian technology of the self is based on the notion of 'government,' and it extracted from the individual a particularly sinister regimen of 'unconditional obedience, uninterrupted self-examination, and exhaustive confession' (Foucault, quoted in Lochrie, 1997, p. 6). This process of producing knowledge about oneself by way of disciplined self-inspection is today incorporated in neoliberal technologies of governance with – and not against – selves, whose agency and autonomy ultimately rest on the capacity of ongoing monitoring and self-regulation.

While the first part of the book inquires into the genealogy of self-technologies producing those individualistic and voluntary selves current neoliberal governmentality so heavily relies upon, the second part of the book is most of all concerned with the systems of power/knowledge giving rise to or maintaining this kind of self. In Chapter 3 on 'Governing the Will in a Neurochemical Age', Nikolas Rose shows how the brain has become the key to the question of human mental life. Whereas psychology held the brain merely as the physical locus of human psyche, Rose identifies four dimensions promoting a shift in human ontology: brain imaging, molecular neuroscience, psychopharmacology, and behavioral genomics. Rose marks this shift 'in the way of seeing, judging and acting upon human normality and abnormality' as dispensing with (or at least supplementing) 'psychological' individuality and converging on what can be called 'somatic' individuality (Novas and Rose, 2000). Whereas initially human beings were conceived of as driven by the brain as an internal space shaped mainly by their biography and experience, today key aspects of their individuality are defined in 'bodily terms'. This tendency allows for operations on the body itself, especially on the brain. Furthermore, it is not the correction of abnormalities, but the management of susceptibilities that is taking center stage. Ultimately, this will not leave the self-technologies mentioned above unaffected: as Rose maintains elsewhere, this shift gives rise to neuroceutical interventions which are not so much about widening the net of pathology. Instead, 'we are seeing an enhancement in our capacities to adjust and readjust our somatic existence according to the exigencies of the life we wish to aspire' – and the self we wish (others) to be (Rose, 2004).

From a different angle, Armin Nassehi in Chapter 4 examines sociology's role in constructing the figure of an actor based on the classical autonomous subject. Whereas contingency and higher ranges of decisions and the ascription of action to a willing or rationally choosing person have required the invention of inner spheres of the persons, Nassehi's chapter, 'The Person as an Effect of Communication', describes sociology as a project deconstructing this theoretical figure, albeit shying away from the necessary consequences of refusing the bourgeois figure of the subject. Although social theories and empirical practices of sociology postulate the actor as the source of social 'energy', they hardly ever reflect on this prerequisite. As great parts of sociology believe in the actor, not mentioning that the actor himself/herself is a point of attribution, sociology treats as a solution what should be its genuine problématique: how the actor becomes an actor altogether and why modern society counts on the free will of individuals. To this end, Nassehi offers a sociological alternative, dealing with this problématique by outlining an inclusion theory. With inclusion theory, he refers to systems theory and one of its basic assumptions: communication as the basic element of social systems. In this perspective, inclusion stands for 'communicative strategies for considering human beings as relevant' and persons emerge as an effect of communication (thus also the title of Nassehi's contribution) as it is communication that renders subjects accountable.

In the third part of the book, the authors scrutinize the political dimension of individualistic and voluntary selves. Starting off with a neoliberal conception of the self, in Chapter 5, John Clarke, Janet Newman and Louise Westmarland explore the role of the conceptions of the consumer in the reform of public services in the United Kingdom. The shift from citizen to consumer is widely understood as emblematic of neoliberalism and seems to embody a set of much wider distinctions such as state/market, rights/contracts, public/private, collectivism/individualism, and ultimately social democratic welfarism/neoliberalism. In their chapter, 'Creating Citizen-Consumers? Public Service Reform and (Un)Willing Selves', Clarke, Newman and Westmarland are highly skeptical about the reliability and usefulness of the citizen/consumer distinction and the shifts it conveys. Drawing on an empirical study on the identification of users of public services, they show that people actually do not grasp their relations to public services within the binary of citizen/consumer deemed so central to contemporary public, political, and political science debate. For this reason, Clarke, Newman and Westmarland argue for an approach that stresses a 'politics of articulation' rather than a 'politics of subjection': they doubt that subjects emerge because they come to recognize themselves in dominant discursive figures; rather that through forms of identification, the 'conduct of conduct' may be accomplished by the regulating effects of practices and relationships. Although the distance between people and designated subjections is not intrinsically political, as it does not mobilize collective action, Clarke, Newman and Westmarland hint at possible political effects in terms of the power of alternatives to the dominant discourse. In this vein, they identify selves unwilling to adapt to the designated subject positions. Consequently, however, one could argue that such acts of repudiation require willing selves: selves willing to adhere or create alternatives to the dominant.

With a study of neoconservative approaches, Barbara Cruikshank turns in Chapter 6 to a policy that can be read as reaction to the mounting success of neoliberal politics. In her chapter, 'Neopolitics: Voluntary Action in the New Regime', Cruikshank identifies political efforts to dispense with autonomous selves by contrasting neoliberal and neoconservative policies in the contemporary United States. Whereas neoliberal policies rely upon the autonomy and economic rationality of the individual will to replace governmental functions, neoconservatives deem the willing self to be an effect of neoliberal governing that is tamed by measures of remoralization. According to neoconservative belief, market rationality without any state intervention cannot produce the moral ground on which it stands, and therefore the neoliberal retreat of state action in favor of the market principle is a fatal undertaking. Neocons claim authoritarian state intervention to be the appropriate remedy against a force produced by neoliberal policy and conceived of as dangerous: free will. It is Cruikshank's contention that the linkage of power–knowledge, so painstakingly described by Michel Foucault, is more or less broken in these efforts to 're-moralize' the state and civil society. Power uncoupled from knowledge does not operate productively according to a norm, but negatively against the pluralization of norms. The revitalization of civil society as the domain of voluntary action is no longer the target of neoconservative reform or the antidote to big government and welfare dependency. Instead, civil society is re-governmentalized in a way that threatens, but may not defeat the strategy of reform.

Mariana Valverde explores the logic of a certain type of governance that dispenses with the idea of governing society as a whole. In contrast to such universal governance, her observations indicate the increasing popularity of what she calls targeted governance. For example, as Valverde has shown elsewhere, policing tends to be targeted geographically or risk-factor specific. In Chapter 7, ' "Craving" Research: Smart Drugs and the Elusiveness of Desire', she explores the idea of targeted governance not in abstract theoretical terms, but in relation to a particular target: alcoholism. As an object that has long been a site for all sorts of discourses and governing techniques aimed at human willpower, the treatment of alcoholism proves an illustrative case for governing willing selves.

Just as smart bombs are supposed to selectively affect particular targets, the development of smart drugs aims at correcting brain disorders unconstrained by the idea of an identity as alcoholic, depressive, or the like. Using 'smart' drugs as an example, Valverde shows that the need for constant assessment of results achieved and its feedback to authorities in order to develop and improve the means of targeted governance indicates that this does not amount to governing less.

The fourth and final part of the book concerns itself with ethics and morality linked to current concepts of the will. Kenneth J. Gergen, in Chapter 8, 'From Voluntary to Relational Action', shows that the concept of voluntary action may legitimately be traced to Aristotle's writings on the psyche. However, history also provides us with numerous cases in which the concept has been appropriated for various purposes and projects. Of major significance today is the way in which the concept of agency functions, first, as a means of social ordering and, second, as a means of according the human being intrinsic value. As Gergen proposes, the uses of agency in social control are deeply problematic, both pragmatically and ideologically. Yet, to dispense with the concept of agency would also lend further weight to the deterministic model of human functioning favored by the sciences (including psychology). In effect, the intrinsic worth of the human being would be destroyed. In this context, Gergen outlines a relational account of human meaning and action. Drawing from post-structural and Wittgensteinian philosophy, this account will essentially deconstruct the traditional voluntarism/determinism binary. Tracing the source of all reality posits, logics, and values to relational process, the attempt will be to replace institutions of individual responsibility (blame) with practices of relational responsibility.

Tillmann Vierkant in Chapter 9 on 'The Role of the Self-Model for Self-Determination' argues for a conception of free will that is based on moral realism. As the idea of free will and responsibility is centrally tied to the notion of autonomous action, Vierkant discerns three self-types that account for differing intuitions about the meaning of 'autonomous'. He presents a conceptual clarification of what remains largely implicit in more sociological accounts on autonomy. Thereafter he comes up with a suggestion according to which one could argue that the 'true self' is indeed causally relevant for individual action. On Vierkant's account, the true self is a set of causal determinants that are established within a social context. Stereotypes a society deems useful (e.g., a stereotypical help response towards children) might constitute such building blocks of the true self. They are not only able to cause behavior (e.g., to help children), but they can also be understood as self-determination. Namely, if we understand not only individuals but also social groups as disposing of reasoning capacities (congealed in, e.g., stereotypes), then it might indeed make sense to speak of individual behavior as being self-determined in the sense that it is caused by socially approved reasons.

Although only a few contributions refer to the concept of a society of control (Deleuze, 1990) directly, they widely cover the changes it implies: they are all about 'forms of free-floating control'. Whereas technologies of discipline are aimed at the fabrication of subjects within enclosed environments, such as family, school, and factory, technologies of control are continuous and ubiquitous throughout society. The latter entail 'liberating and enslaving forces' confronting one another, thereby guiding individual action and rendering it socially highly predictable. The willing self that is enunciated within and in-between such processes is produced, yet not fully determined by those highly variegated technologies of control.

## On willing selves *vis-à-vis* the challenge of the neurosciences

Heterogeneity and internal differences notwithstanding, all contributions in this book ultimately subscribe to the idea that much has been, and still is being done (socially, scientifically, technically, philosophically) to bring the autonomous, willing self about. Albeit sometimes unconsciously or uncritically, most societal practices now involve, if not co-produce, the individual as responsible, prudent, and skilled, ready to be included into governmental practices. Governed by individualistic and voluntaristic practices, to be sure, a neosocial self is about to emerge – feeling and being held responsible for itself and society.

The remainder of this chapter, addresses the question why it stands to reason that the neurosciences should declare the will an illusion just when neoliberal society sets out to hail the agentive, autonomous, voluntaristic self. While the neurosciences as such aim to liberate us from misconceptions inconsistent with modern scientific findings, the picture changes dramatically once the society of which the neurosciences are part is considered: it is neoliberal society – a society currently debating a more active, enterprising, and responsible role of its members. Hence, by debating the (non-) existence of free will, the neurosciences – if unknowingly – become a stage for public deliberation of modes and rules of sociability: individual agency, responsibility, and guilt frame the debate on the possibility of sociality in neoliberal society. Neoliberal society, thus Stephan Lessenich's reply, is at the same time a neosocial society. The reason being that it 'constitutes itself as a subject that demands active citizenship. Society is now prime reference of sociality and evaluates individual activities according to their degree of sociality' (Lessenich, 2003, p. 89). This requires the individual capacity to monitor and control oneself – for the benefit of oneself and society. In this perspective, we suggest slightly adapting the terminology, as we prefer to talk about *neosocial selves* being constitutive of *neoliberal societies.*

Thus, the coincidence of the neurosciences declaring the will an illusion while neoliberal society sets out to hail the agentive, autonomous, voluntaristic self is, in our view in need of further qualification in three respects.

*Qualification 1.* Although neuroscientific accounts and social practices in neoliberal society may seem contradictory at first, they are, in fact, not, as they simply refer to different levels. At one level, physical claims are made, such as in the neurophysical narrative about how the brain regulates what happens 'inside'. This is a mechanistic and deterministic account of our (neuro-) biological bodies. At another level, a psychological narrative about ourselves as persons is told. At this level, actions performed by persons are described, reasons for these actions given, and questions addressed as to what this person thinks or believes, and so on. Freedom can only be thought of, denied, or be deliberated at this very level of personhood. Attempting to do so at any other level is a mistake in category (Bieri, 2005).

*Qualification 2.* Neuroscientific accounts and social practices are also non-contradictory in a second respect. For social practices, when attributing free will, autonomy, or consciousness, no longer refer to 'traits' characteristic of a 'mature personality.' Rather, by appealing to voluntariness an individual's capacity for governing itself and others is addressed. That is, sociability today emerges from practices attributing voluntary action with agency and responsibility (Maasen, 2000a, b). Likewise, training our capacity to decide, exerting our 'free will,' is not so much about enhancing our personality as about improving our functionality as sociable selves.

*Qualification 3.* Finally, given the aforementioned, it is not the neurosciences that re-shape society according to their findings. Rather, so it seems to us, the neurosciences are the sciences conforming to neoliberal society hailing neosocial selves. Particularly with regard to the recent boom of neuroceuticals, one may easily infer how pharmacological intervention aligns with a neoliberal regime of individuals who must govern themselves and who do so by employing self-management practices of various kinds. That is, neuroscientific discourse contributes to rearranging the roles and interrelations of free will, selves, and society – not voluntarily but *willy-nilly*.

For all these reasons, the neurosciences will not be the subject matter of this book, but rather figure as one prominent example of current challenges to self, society, and their relations. They are – as a science, as a technology, as a subject of public hope and concern – a prototypical example of developments in current society. And it is these the authors of this book are primarily interested in. The buzzwords are: free will, (neosocial) selves, and neoliberal society.

True, it is a commonplace that 'autonomy' has assumed a certain polysemy in contemporary moral and political discussion. As Stephen Darwall points out, the term's original meaning was strictly political: a right assumed by states to administer their own affairs, and it was not until the nineteenth century that 'autonomy' (in English) came to refer also to the conduct of individuals (Darwall, 2006, p. 263), though the meaning differed from current use. Today, its meaning is tightly connected to both the individual and the political, both of which are tightly connected by the autonomous self – not least due to the regime of neoliberalism.

## References

Bieri, P., 'Untergräbt die Regie des Gehirns die Freiheit des Willens?', Manuscript (2005).

Bröckling, U., S. Krasmann and T. Lemke, *Gouvernementalität der Gegenwart. Studien zur Ökonomisierung des Sozialen* (Frankfurt/M.: Suhrkamp, 2000).

Crick, F. and C. Koch, 'Consciousness and Neuroscience', *Cerebral Cortex*, 8, 2 (1998): 97–107.

Darwall, S., 'The Value of Autonomy and Autonomy of the Will', *Ethics*, 116 (2006): 263–284.

Dean, M., *Critical and Effective Histories: Foucault's Methods and Historical Sociology* (London and New York: Routledge, 1994).

Dean, M., 'A Social Structure of Many Souls?: Moral Regulation, Self-Formation and Government', *Canadian Journal of Sociology*, 19 (1995): 145–68.

Deleuze, G., 'Society of Control' (1990). Accessed 17 July 2006 at: http://www.nadir.org/nadir/archiv/netzkritik/societyofcontrol.html

Fitzsimons, P., *Radical Pedagogy, Neoliberalism and Education: The Autonomous Chooser* (2002). Accessed 5 January 2007 at: http://radicalpedagogy.icaap.org/content/issue4_2/04_fitzsimons.html

Foucault, M., 'The Political Technologies of Individuals', in L. H. Martin, H. Gutman, and P. H. Hutton (eds), *Technologies of the Self* (Amherst: University of Massachusetts Press, 1988), pp. 145–62.

Foucault, M., 'Questions of Method', in G. Burchell, C. Gordon and P. Miller (eds), *The Foucault Effect: Studies in Governmentality* (Chicago: University of Chicago Press, 1991), pp. 73–86.

Fowler, G. L., 'The Human Genome Project – what's the public got to do with it?', *Public Understanding of Science*, 8, 3 (1999): 153.

Gordon, C., 'Governmental Rationality: An Introduction', in G. Burchell, C. Gordon and P. Miller (eds), *The Foucault Effect: Studies in Governmentality* (Chicago: University of Chicago Press, 1991), pp. 1–51.

Hilgartner, S., *Science on Stage: Expert Advice as Public Drama* (Stanford, CA: Stanford University Press, 2000).

Keat, R. and N. Abercrombie (eds), *Enterprise Culture: The International Library of Sociology* (London: Routledge, 1991), pp. 1–23.

Lessenich, S., 'Soziale Subjektivität. Die neue Regierung der Gesellschaft', *Mittelweg 36*, 4 (2003): 80–93.

Lochrie, K., 'Desiring Foucault', *The Journal of Medieval and Early Modern Studies*, 27, 1 (1997): 3–16.

Maasen, S. 'Neurosociety Ahead? Debating Free Will in the Media', in S. Pocket, W. P. Baks and S. Gallagher (eds), *Does Consciousness Cause Behavior?* (Cambridge, MA: MIT Press, 2006a), p. 339–59.

Maasen, S., 'Gibt es den freien Willen? Eine Debatte im Feuilleton', *kultuRRevolution*, 50 (2006b): 44–59.

*Meeting of Minds: European Citizen Jury on Brain Research*, Deutsches Hygiene-Museum, Dresden (25–27 November 2005).

Miller, S., 'Public understanding of science at the crossroads,' *Public Understanding of Science*, 10 (2001): 115–20.

Milner, A. D., 'Cerebral correlates of visual awareness', *Neuropsychologia*, 33 (1995): 1117–30.

Novas, C. and N. Rose, 'Genetic Risk and the Birth of the Somatic Individual', *Economy and Society*, 29, 4 (2000): 485–513.

Rimke, H. M., 'Governing citizens through self-help literature', *Cultural Studies*, 14, 1 (2000): 61–78.

Roth, G., *Fühlen, Denken, Handeln. Wie das Gehirn unser Verhalten steuert* (Frankfurt/M.: Suhrkamp, 2003).

Rose, N., 'Becoming Neurochemical Selves', in N. Stehr (ed), *Biotechnology, Commerce And Civil Society* (Transaction Press, 2004), retrieved 2 February 2006 from: http://www.lse.ac.uk/collections/sociology/pdf/Rose-Becoming NeurochemicalSelves.pdf

Schnabel, U., 'Der Markt der Neuronen. Hirnforscher werden zu Werbefachleuten. Sie wollen enthüllen, was den Käufer zum Konsum treibt', *Die Zeit*, 41 (2003): 47.

Singer, W., *Der Beobachter im Hirn. Essays zur Hirnforschung* (Frankfurt/M.: Suhrkamp, 2002).

Zeman, A., 'Consciousness', *Brain*, 24 (2001): 1263–89.

# Part I
# Self – Past and Present

## Introduction

The first part of this book is devoted to cultural evolution or, as Foucauldians would have it, to the genealogy of recent practices that have become known as self-technologies. Technologies of the self such as therapy, counseling, and self-help (Chapter 1: Sabine Maasen, Barbara Sutter and Stefanie Duttweiler) are prime examples of self-technologies. They are characterized by the various 'operations on their own bodies and souls, thoughts, conduct, and way of being' that people make either by themselves or with the help of others in order to transform themselves to reach a 'state of happiness, purity, wisdom, perfection, or immortality' (Foucault, 1988, p. 18). Foucault deliberately chose the term genealogy to evoke Nietzsche's genealogy of morals as it highlights complex or nonspectacular origins, and to refer to the fact that utterly mundane practices are part and parcel of what we conceive as particularly significant achievements.

In this vein, our modern understanding of ourselves as being highly individualistic and voluntarily acting selves does have a complex, discontinuous, and partly inconspicuous history. As to the latter, the following two chapters probe into the culturally evolved tradition of linking selves and society by way of practices that seem utterly self-centered. A rich body of literature on self-help, but also on the foundational practice of confession, testifies to the paramount importance of self-technologies in bringing about willing selves that today are capable of competently governing themselves and others (family, neighbors, communities). Current projects such as 'values education', 'moral education', 'citizenship education', 'personal and social education', and so on, have emerged, none of which would be intelligible without these highly practiced techniques of telling the truth in order to adequately govern oneself and/or others. Likewise, a mass society would not be possible without its members' capacity to do so. Notably, our current society of control can, and indeed does, rely on this capacity and the self-evidence of constantly monitoring and regulating oneself. The very evidence of this activity is, for a large part, rooted in a network of practices and institutions based on such self-technologies.

However, it is also firmly rooted in a long cultural tradition of relating selves to their society. 'Contemporary notions of confession are derived not simply from the influence of the Catholic Church and its strategies for confessing one's sins (where sin is mostly equated with sexual morality so that confession became the principal technology for managing the sexual lives of believers), but from ancient, pre-Christian philosophical notions.' Furthermore, they have also been 'profoundly influenced by confessional techniques embodied in Puritan notions of the self and its relation to God, and by Romantic, Rousseauian notions of the self' (Besley, 2005, p. 83). Ever since, the confessional practices have been part of the secular world: they have been integrated into medical and then into therapeutic and pedagogical models in contemporary secular societies.

In their various forms and fashions, self-technologies are systems of power/knowledge. 'At least for the study of human beings,' power and knowledge 'cannot be separated: in knowing we control and in controlling we know' (Gutting, 2003, p. 6). Modern societies of control embed this mechanism in most intricate ways. More specifically, control is achieved largely by the internal monitoring of those controlled. Indeed, the current self-helpers must be skilled in their own subjection, in organizing and sustaining themselves as calculable, classifiable, responsible, self-regulating, and, hence, 'governable'. This notion and practice, far from being invented only just recently, is but the latest transformation of techniques of self-thematization (Hahn). It is based upon, and, indeed, is itself a 'technology of the will'.

Namely, as Alois Hahn and Marén Schorch show in minute detail in Chapter 2, the current notion of free will and responsibility cannot be disconnected from its Christian roots, notably not from its core practice of confession. In the confessional, the sinner was led to speak about their (sinning) self – his or her deeds, words, and thoughts. All sins had to be confessed and (shamefully) repented. The confession makes sinners speak about themselves – it is a generator of self-thematization, out of which a (willing) self emerges.

While leaving the reconstruction of the confession and its modern successors (autobiographies and other self-generators) to the authors, we would like to stress the multitude of ways in which confession has indeed operated as a technology of the will. To begin with, sins have to be conceived as deeds, words, or thoughts directed against God's will. Confronting God with one's rebellious will is, in a way, the sin of all sins. Some theories of sin even insist on the fact that it is the very act of agreeing-to-sin, that is, the will to sin, which deserves damnation in the first place. Yet, in order to confess, one has to overcome one's feelings of shame and guilt: the very act of confession proves a strength of will by way of submitting oneself to God's Law. The more the notion of sin enters the domain of inner motives, fantasies, and unruly thoughts, the more the minute workings of the will have to be brought to the surface by way of willful inspection: deeds, words, and thoughts may both indicate and provoke one's will to sin. When the Reformation increased the frequency and ubiquity of these self-inspections, the confession towards oneself, one's wife or husband, or spiritual guide became an ongoing activity. One had to scrutinize not only every possible way in which one may have sinned, but also scrutinize one's (sinful) biography, emerging from it in order to learn about one's chances for Eternal Salvation.

To summarize: confession, in its Christian as well as its modern variants, is a prime technology of the will. The technique is to make the sinner speak, that is, to induce the will to confess his or her sins that, in turn, are regarded as willfully enacted deeds, words, or thoughts against God's will. To be sure, this did not occur without guidance: so-called penitentials (summae confessorum) instructed both the priest and the sinner on how and what exactly to confess (Maasen, 1998). For instance, in 1867, Gaume issued a penitential in late medieval style. Before explaining the questions the priest was to ask, a third of the manual is devoted to his tasks before, during, and after the act of confession. The first chapter introduces the priest to his different roles throughout the process. Acting as a father, doctor, teacher, and judge, the priest's foremost duty is to evoke a truthful confession. The manual even offers introductory remarks such as, 'Say everything without hesitation and do not be ashamed. It does not matter that you did not examine yourself sufficiently. It will suffice if you answer to my questions … You should have no doubt, God will excuse you, if you have good intentions …' (Gaume, 1867, p. 67). The manual thus provides the priest with a pedagogic program, explaining his tasks, teaching him by examples, and giving meticulous instructions on how to behave in different situations. In the following chapters, there are detailed instructions that elaborate on the actual sins that may have possibly been committed. These instructions are designed to guide the priest in his attempt to elicit a truthful and complete confession.

In this way, calling confession a technology is more than just a metaphor: rather, the term 'technology' highlights the systematic character of the Christian discourse on willing selves. It defines a problem (sins against God's will), outlines its dimensions (sins of thoughts, words, and deeds), offers a way to its solution (willful confession), and specifies sub-goals (shame, remorse, penance) which help to willfully control the state of perfection. Thus, viewing confession as a technology allows us to see the will emerging as both a vehicle and the target of systematic self-inspection, producing a self-conscious and responsible self.

The rise of self-help should therefore not be considered a transitory cultural fad. It should rather be seen as a correlative of practices and techniques based on neoliberal governmentality. It is a mode of rule that signifies the emerging audit culture (Powers) that increasingly exerts control by means of, not against, individuals. It both disciplines and makes use of their agency, ultimately, by way of self-technologies.

## References

Besley, T., 'Foucault, truth telling and technologies of the self in schools', *Journal of Educational Enquiry*, 6, 1 (2005): 76–89.

Foucault, M., 'Technologies of the self', in L. H. Martin, H. Gutman, and P. H. Hutton (eds), *Technologies of the Self* (Amherst: University of Massachusetts Press, 1988), pp. 16–49.

Gaume, J., *Handbuch für Beichtväter* (Regensburg: Georg Josef Manz, 1867).

Gutting, G., 'Michel Foucault', in *Stanford Encyclopedia of Philosophy* (2003). Accessed 5 January 2007 at: http://plato.stanford.edu/entries/foucault/.

Maasen, S., *Genealogie der Unmoral. Zur Therapeutisierung sexueller Selbste* (Frankfurt: Suhrkamp, 1998).

# 1
# Self-Help: The Making of Neosocial Selves in Neoliberal Society

*Sabine Maasen, Barbara Sutter and Stefanie Duttweiler*

## 1 Self-help books: governing oneself and others by will

Throughout the last two decades, therapeutic practices have come to permeate society in various forms and fashions. Indeed, various authors have noted that we live in a world that may justly be characterized by an intensified and highly variegated preoccupation with the self (e.g., Giddens, 1991; Maasen, 1998; Rimke 2000, p. 61; Taylor, 1989). In addition to therapy proper, counseling and self-help have assumed increasing popularity. Journals, radio, television, Internet – wherever one looks, one finds yet another new version of (one-way) therapy and counseling. While the individual variants differ enormously as to their ambition, expertise, and impact, they all imply certain requirements in terms of a highly specific kind of communication: first, they are firmly based upon everybody's capability to perform a demanding discourse called therapeutic communication. It entails our ability to present a problem to an expert, who will then help us, as a layperson, to solve that problem in various settings (be it short-term, long-term therapy or counseling). Second, they require that one knows when to seek professional help and how to choose among various offers on the market (you pick up the phone, click into the self-help chat, or buy a book). Third, in all cases you need to transfer the lessons learned in special settings (on the couch or in a group seminar) into your everyday life.

Self-help by 'how-to' books has introduced yet another significant modification to these requirements: it is a solitary practice of buying and reading a book, and then processing the advice given, or not. You yourself are the expert with whom you have to reach an agreement and who ensures that an agreement is met. Diagnosis, goal setting, and change all take place in one person who is simultaneously both therapist and client. The medium of self-help books thus constitutes a most intimate setting of self-change. And there is no denying that modern selves increasingly possess all the aforementioned skills: 'self-help literature has become an enduring, highly fashionable non-fiction genre, especially within the last twenty five years' (Rimke, 2000, p. 62).

In the following article, we treat this literature as a practice that is both based upon and produces the notion of willing selves. Indeed, self-help is an activity presumed to be individualistic and voluntary. Relying upon notions such as choice, autonomy, and freedom, self-help furthers the principle of individuality, entailing self-modification and 'enhancement'. This becomes most obvious when succumbing to their promise: change is possible; you just have to want it! To this end, the will, *your will*, assumes two tasks. First, the will is the initiator for all measures taken to educate and discipline yourself. Second, the will is an important *vehicle* and therefore a prime *target* of all those measures taken to educate and discipline yourself. To state it bluntly: in the course of your self-managing efforts, you have to find out what you really want. For only what you really want you do (Sprenger, 1997, p. 71). Note the reverse: '*Who says "I can't" really means, I won't*' (Sprenger, 1997, p. 30).

In so doing, self-help literature proceeds on the assumption that people can exercise control over their lives. Interestingly enough, this postulate of self-mastery is highly consistent with the political rationality currently hailed in neoliberal democracies. Indeed, self-help literature proves particularly instrumental in the production, organization, dissemination, and implementation of particular liberal modes of truth about the social world, the individual playing a key role. As Heidi Marie Rimke puts it, self-help is 'an individualized voluntary enterprise' to modify or transform the self. It 'presents individual development as a free moral and ethical decision and as a "natural" undertaking' (Rimke, 2000, p. 63). Rather than being perceived as an authoritarian or subjugating move, this practice is embraced by individuals who thereby come to value themselves and others as responsible citizens – responsible towards themselves and others (family, neighbors, employer, community, society), that is.

In the terminology of Michel Foucault, the practice of self-help is a self-technology. Self-technologies 'permit individuals to effect by their own means or with the help of others a certain number of operations on their own bodies and souls, thoughts, conduct, and ways of being, so as to transform themselves in order to attain a certain state of happiness, purity, wisdom, perfection, or immortality' (Foucault, 1988, p. 18). Granted, advice books regularly aim at very mundane goals: success, health, youth, beauty, happiness, or at active citizenship, respectively, yet those of us who desire guidance are provided with minute instructions referring to our soul, our behavior, our whole form of existence – hence, to our selves. The books proceed by establishing a relationship with oneself, the basic 'operations' being countless techniques of self-reflection and exercises to empower oneself. How-to books, so it seems, are sites and media of thorough exchange with oneself.

The current self-help hype, however, has not come out of the blue but is a societal necessity in an era of frequent change and uncertainty; hence the plethora of therapeutic and counseling offers. Most introductory chapters in self-help books convey the message that while we cannot change the world, we can – or even actually *have to* – change ourselves. Indeed, self-change work is the buzzword of our times. Is this a promise or a threat? Given the fact that advice books are not only bestsellers in every bookshop at present, but have also accompanied the evolution of Western societies through all stages of differentiation and the concomitant shake up of relations between individuals and society, there is no easy answer at hand. It seems as if advice books accompanied and also co-constructed time-bound notions of what a self is and how it is capable of adjusting to and governing his or her life in society. They do so by naming 'problems,' giving 'advice' as to possible answers that may or may not lead to a novel 'behavioral response' in the reader. To this end, advice books address problems at the level of knowledge. They distribute, evaluate, and operationalize knowledge framed as advice on what (not) to do. They make use of all kinds of knowledge (e.g., scientific, ethical, therapeutic) and of various forms of presenting it (e.g., prose, maxims, photos, graphs). They are normative, yet increasingly, their normativity is confined to what gave them their generic title: they insist on *how to* rather than on *what* to do. What is important is the fact *that* one changes, for this is the way to adapt to inner and outer circumstances and, hence, the *via regia* to becoming happy, healthy, successful. In a society of permanent change, how-to books, so it seems, are sites and media of individually responsible responses to these environmental turbulences.

Self-help thus creates 'healthy' citizens inasmuch as they become skilled in organizing and sustaining themselves as an 'operative unity': That is to say, intelligible selfhood today is not about a fixed identity but about a flexible self, always capable of responding to inner and outer challenges. 'By marshalling the concept of responsibility, popular self-help discourses provide an example of how the operations of power in everyday life can incite governance of the self thanks to expert pronouncements about both success and morality' (Rimke, 2000, p. 63). Those who render themselves predictable, calculable, classifiable, self-regulating, and self-determined, hence 'governable,' may justly be regarded as responsible citizens: *voilà*, the social subject of neoliberal governance.

In this perspective, changing oneself by intensive exchange with oneself at all times, particularly at stages of individual and/or societal change is a program that thus deeply affects the notion of sociality. For free will and responsibility are embedded in a debate about the possibility of sociality today: other than as authors such as Rimke would have it (2000, p. 67), modern selves are not urged to become non-social. Rather, the modern self-regulated self is constituted as one that is responsible towards both itself and society. Neoliberal societies are at the same time neosocial societies. Neosocial society, in the words of Stephan Lessenich, 'constitutes itself as a subject that demands active citizenship. Society is now prime reference of sociality and evaluates individual activities according to their degree of sociality' (Lessenich, 2003, p. 89). This requires the individual's capacity to monitor and control themself – for the benefit of themself and society. How-to books represent a prescriptive kind of literature that testifies to the time-bound linkages between society and selves – the novel arrangement emerging in neoliberal society is one that produces neosocial selves. They are characterized by their capacity for ongoing self-regulation which promises operative stability and structural flexibility alike: whereas the latter allows for permanent adaption to changing circumstances, the former counters the risk of succumbing to the turbulences by way of governing oneself according to accepted forms (e.g., self-help) and norms of action (e.g., help yourself).

More precisely, our chapter is guided by two basic assumptions. First, practices of self-constitution designate the subject's prime mode of governance. This mode engenders a specific type of subjectivity: it is structurally unstable, yet operatively stable. Although always a 'work in progress' and continuously involved in (self-) reform, the novel mode of subjectivity is stable with respect to its practices of self-reform. The techniques and practices applied (such as self-help) are socially distributed and provide mutually expectable means of adapting oneself. The second assumption refers to the new mode of sociality within the neoliberal rationality. We hold it an important aspect of neoliberal society that it gives rise to selves that are structurally flexible and adaptive, yet are so by operating in mutually forseeable modes of conduct. Being neosocial is thus tantamount to individuals that flexibly govern themselves and others by way of socially accepted means.

To substantiate this claim, the forthcoming study will first address the genre of self help-books more closely: it scrutinizes their level and mode of operation as well as their history of linking selves and society (Section 2).

It is designed as a brief survey of the prototypical ways in which time-bound notions of what a self is and how it is capable of adjusting to and governing his or her life in society have been constructed in and by this literature (Section 3). After summarizing the main techniques for governing selves (Section 4), we will then proceed to educational handbooks for active citizenship, thereby revealing this genre as a technology of neosocial selves in neoliberal societies (Section 5). Self-help, in other words, is both a socialized and sociable solitary practice. Today, this practice finds itself among the acceptable forms of (self-) government and gives rise to novel ways of conducting moral and public discourse, combining discipline and control (Section 6).

## 2 Self-help: knowledge for orientation

Advice regularly operates at the level of knowledge. Typically, it is knowledge for orientation. 'This book will be one of the most important books in your life' (Küstenmacher, 2001, p. 11). A major feature of our society, which has come to be known as a knowledge society, is the production and use of knowledge designed to cope with ongoing change, complexity, risk, and uncertainty in various domains (e.g., education, economy, politics). Modern societies are not only based on knowledge but are also, possibly, threatened by the multiplication of knowledge. Recent times have witnessed the emergence of a plethora of types and procedures of knowledge that organize, steer, and guide the production, dissemination, and evaluation of knowledge. These types and procedures may be seen as second-order knowledge designed to counter dysfunctional tendencies in the knowledge society (e.g., disorientation and incapacity to act). Hence they accomplish what may be called orientation, or, more precisely, governance by knowledge. In political arenas, for instance, we find a growing body of professional advice as well as arenas of ethical deliberation that engage in structuring and mediating different forms of knowledge and values so as to guide individuals or advise political bodies. In the area of the individual conduct of life, we find all kinds of educational activities (life-long learning) as well as therapy, counseling, and self-help.

The fact that expert knowledge is called for in all domains should be observed in regard to a shift of the ambits deemed as amenable to political deliberations. 'How might one reconcile the principle that the domain of the political must be restricted, with the recognition of the vital political implications of formally private activities?' (Rose and Miller, 1992, p. 187). Following Rose and Miller, this is exactly the task expertise is thought to accomplish: 'Experts hold out the hope that problems of regulation can remove themselves from the disputed terrain of politics and relocate onto the tranquil yet seductive territory of truth' (Rose and Miller, 1992, p. 188).

On the level of knowledge, recent self-help can be regarded as a hybrid genre: whichever exemplar of how-to book one chooses, it abounds with all kinds of techniques designed to guide one's reflection about oneself and to change one's behavior and attitudes. The techniques are drawn from a heterogeneous set of sources: psychology (questionnaires, motivation aids), physiology (diet plans, fitness schemes), philosophy (art of living, how to become happy), common sense (proverbs), the esoteric (spiritual knowledge), and wisdom (from philosophy, religion, or everyday knowledge alike; Duttweiler, 2006). As 'practices by which the subject is defined and transformed are accompanied by the formation of certain types of knowledge ... organized around forms and norms that are more or less scientific', Foucault identifies an obligation for any subject, namely 'to know oneself, to tell the truth about oneself, and to constitute oneself as an object of knowledge both for other people and for oneself' (Foucault, 1997, p. 177). Becoming an 'object of knowledge' in this respect is a prerequisite for becoming a 'subject/object of governance' (Rimke, 2000, p. 68). With the knowledge provided by self-help this is processed both under 'the gaze of an expert acting at a distance' but also, and ever-more importantly, under 'the ever-present gaze of one's self' (Rimke, 2000, p. 68). Expertise thus comes in various forms and fashions; expertise in self-help most clearly exhibits the two-faced character of any kind of expertise: it is a technology of freedom and control alike (see below).

While most self-help books favor a certain set of knowledge types and their concomitant set of techniques, they rarely refrain from providing their readers with some proto-sociology. Rather than immersing the subject in their micro-worlds, most books offer their reader, lessons into the world-as-it-is. They thereby provide the helpless with information not only about the society surrounding them, but also about how society and the self relate. Hence, advice books are not only about selves, but also about selves and others, and, ultimately, about selves-in-society. From a sociology of knowledge point of view, how-to books allow studies in selves and society alike. Conventionalizing the measures to be taken by the individuals when making choices, how-to books are media of individualization and, at the same time, media of normalization. In other words, they are media of making neosocial selves.

Before introducing the most prominent kinds of techniques, a brief account will show how this increasing focus upon working on oneself has evolved over time. Rather than a historical study of this genre, it is a genealogical account that proceeds from the current obligation to work on oneself and tracks the specific evolution of this seemingly novel obligation. Thus we focus on prototypes of self-help books, particularly on those that are constitutive for the emergence of this particular practice. This is essentially a study of the increasing frequency and intensity of working on oneself, always specifically adapting one's relation to society and its demands (i.e., success). In the course of this happening, self-help has come to be an important medium for producing and disseminating the knowledge required to shape this relation.

## 3 Self-help: a brief study of a long success story

Advice books are not a new phenomenon.[1] From the fifteenth century onward, manuals designed to educate us in manners and virtues have accompanied the modernization of Western societies: according to David Riesman, these manuals were designed to produce character types guided by tradition, rather than individual types we have today (Riesman, 1950). While this notion has become subject to discussion, it can be said that etiquette books are safeguards of forms, manners, and ceremonies established by convention that, in turn, are guided by nothing else but the (Godly) order. As a whole, this literature is canonical and strictly oriented towards rules. The relation with oneself is a 'must' to solve the problems of interaction and only rarely is it a pleasure – with the exception of carefully designed intrigues, that is. Generally, however, the readers should know the rules and follow them, the ultimate goal being to prevent disgrace. As to the individual seeking advice, etiquette books, to be sure, are not about looking for uniqueness but about uniformity. It is precisely this necessity, however, that initiates and enforces an increased attention to oneself in relation to ever more concrete situations. A few examples will illustrate how we learned the necessity to look at ourselves and developed an outright desire to do so.

In the course of the eighteenth century, along with the decline of stratified society, strict societal codes for communication and behavior were eroded. With the emergence of bourgeois society, a different kind of advice was needed for a different kind of readership (though paralleling the older form). Most of all, educated laymen sought advice on how to cope with everyday problems in all dimensions of life. The *Not- und Hilfsbuechlein fuer Bauersleute*, written by the educator Rudolf Zacharias Becker (1787), for instance, tries to 'redress the distinguished bodily and spiritual necessities of the countryman' on about 800 pages. He expands on moral principles, gives 'practical hints,' tries to counter superstition, and even explains how to revive hanged and frostbitten persons or those struck by lightning. Becker's teachings on a so-called order of life for the healthy, the sick, and the convalescent are somewhat closer to modern concerns. The book *Ueber den Umgang mit Menschen* by Freiherr von Knigge (1788) is another, well-known example of a bourgeois advice-book. While it is about learning the rules of behaving in public and in the private sphere and is thus similar to the aforementioned type of books, it introduces a significant shift. The very first chapter of Knigge's book is about 'how to associate with oneself'. The author argues that our duties toward ourselves are the most important ones, and indeed the association with oneself is neither the least useful nor the least interesting (Knigge, 1788, p. 82). In addition to the call 'Be yourself a pleasant company!' made against the danger of idleness, Knigge appeals to truthfulness, steadiness, and a belief in austerity toward oneself: in lonely hours, one should take stock of one's behavior and attitudes. Acting upon oneself as a severe judge, one should sincerely assess whether and how one has made use of the hints given in his advice-book so as to gain perfection (Knigge, 1788, p. 86). These pieces of advice are located in the context of enlightenment and clearly entail an educational intention: a prime vehicle of (self-)education is scrutinizing oneself.

These books address everyday or professional behavior, thereby indicating increasing social mobility and accompanying it with good advice. The books show and propagate a modernization of manners, as well as an increasing difference between public and private behavior. The very moment individuals emerged as unmistakably unique entities, tension arose between themselves and their society calling for a rationality of 'how to cope with situation $x$'. Since then, this rationality has marked the concept of modern individuals, as well as the relation of self and society. Selves, while becoming increasingly more complex, are now equipped with pragmatic rules designed to cope with each other and with themselves (Pittrof, 1989, p. 15).

In the 1920s, we observe the shift toward a more dynamic and goal-oriented type of advice book, resulting from the increasing complexity and contingency of modern society. While social mobility had its own needs, it created possibilities by the same token, thus reorientating the goals of instruction books. In the 1920s, these books focused on the goal of success in the work sphere and how to proceed methodically in order to reach this goal. These advice books focus on developing a goal-oriented personality and making efforts to attain one's goals. Potentially, all humans have the capacity to educate themselves – with the help of books known as 'schools educating the will' (*Willensschulen*), or guides to success.

The book titles of the schools educating the will signal the need for method and enthusiasm alike: *The gymnastics of the will: Practical instructions to enhance energy and self-control, invigoration of memory and pleasure to work by reinforcement of willpower without outside help* (Gerling, 1920), *School for the will* (Lindworsky, 1927), *Life according to the art of generalship* (Sartorius, 1929), *Power and action: A guide to will, health, and power* (Helmel, 1928). While all books focus on willpower, so-called guides to success differ from the schools for the will in one crucial aspect: in their view, there is no such thing as 'a will' but there are only exercises of will performed using specific objects in order to form special competences and to satisfy concrete needs (Grossmann, 1927, p. 164). They instruct their reader how to commit oneself exactly to certain kinds of (realistic) goals and to reach these goals with the help of how-to plans and self-motivation.

In order to gain the correct relation to oneself, one has to proceed in a disciplined way. In every situation, we see rational means that must be applied precisely, methodically, and continuously and to all matters of organizing one's personal life. One has to steel one's will, to train one's body, to educate and control one's mind, the ultimate goal being economic success. From the early twentieth century onward, selves have to differentiate as well as integrate the work on themselves: key variables are the methods of self-energizing, self-education, planned, effective action, self-enthusiasm and endurance. Ever-new hindrances have to be overcome by ever-new decisions (Lindworsky, 1927, p. 56) in order to defeat the dreaded opposites called 'weakness of the will' and 'nervousness'. Working on oneself has become incessant.

At the end of the 1930s, a novel variant of success books emerges, Dale Carnegie's manual *How to Win Friends and Influence People* (Carnegie, 1938) being the prototype. In this literature, the self's dependency on others is the pivot of success. Only those who know how to use others will become successful. It urges the reader to take the perspective of the other: 'If out of reading this book you get just one thing: an increased tendency to think always of the other person's point of view, and see things from his angle ...' (Carnegie, 1938, p. 63). This move, however, is not about internalizing the generalized other in order to produce a 'me' (Mead). Rather it is about coping with the concrete other in order to reach one's own goals. With empathy and flattery, one will gain the other's appreciation and help. Riesman declares Carnegie's book to be an outright textbook of other-directed social interaction. 'It is apparent enough that the guides to attractiveness, living, and success ... are promoters of other-direction. Their message is one that implies constant need of approval by others' (Riesman, 1950, p. 217). The presented strategies can be named as 'other enhancement': help others, give them what they want, agree with people. The type of rationality that informs the Carnegie kind of success books is one of manipulation. It marks both the relation to others and to oneself.

Yet another shift occurs in the 1960s and 1970s when a more psychotherapeutic-oriented self-help literature appears. Following Humanistic Psychology and the Human Potential Movement, the focus is less on discipline and more on 'growth,' 'unfolding,' and 'self-realization'. The oppressed individual liberates itself from social pressure. 'Rich or poor, black or white, male ... we share a belief that feelings are sacred and salvation lies in self-esteem, that happiness is the ultimate goal and psychological healing the means' (Moskowitz, 2001, p. 1). The techniques are well-known in the self-help universe: 'employ liberal self-praise; recall positive events in your life; try some positive hypnosis; learn to listen to yourself; don't punish yourself; learn to accept yourself; make friends with yourself; and so on' (Starker 2002, p. 121). While de-emphasizing discipline, caring for one's growth could not deny a certain amount of obsession: the premise that 'human potential is unfulfilled' leads to a plethora of activities designed to recognize and realize inner needs and feelings. Liberation from inner restrictions and outer repressions should eventually reveal the 'true self'.

The ostentatious obligation to grow in order to be happy produces to the 'libertarian paradox: the sole social obligation is in its negation' (Koch-Linde, 1984, p. 56). As the author of the bestseller 'Your Erroneous Zones' bluntly remarks: 'state your Declaration of Independence from their control' (Dyer 1976, p. 136, quoted in Koch-Linde, 1984, pp. 47–8). Freedom from social pressure goes hand in hand with self-realization in order to gain responsibility toward oneself. Again, another bestseller by Newman and Berkowitz (1971) puts it succinctly: 'How to Be Your Own Best Friend'. Important building bricks of these and other popular forms of self-help were the devaluation of the other person and the upgrading of oneself, the striving for individual happiness, and the reduction of social obligations (Koch-Linde, 1984, p. 57). Towards the end of the 1980s, this development came to a peak with the enforcement of a tendency to regard all kinds of other-relations with suspicion, the catchword being 'co-dependence' (Greenberg, 1994).

The relation one is asked to establish is focused upon oneself. It is all about disclosing and articulating one's true and authentic self. Yet this search for one's true self is fraught with a paradox: is this my true core identity, my real need, my authentic desire? This paradox of authenticity is processed by continuously working on oneself (Trilling, 1982). Each doubt provokes new activities. The process of revealing one's true self is insoluble. Accordingly, working at a truthful and reliable relationship with oneself is a never-ending process.

In the 1990s, the call for 'self-management' changes the relationship with oneself yet again. Self-management no longer promotes the quest for the true self, but subjugates it to the quest for efficiency. Consequently, manuals hail the flexible self that operates best upon a flexible, self-organizing mode. Management is not confined to economic success but rather becomes the dominant metaphor for the organization of all domains of one's life (home, office, friends, hobbies, etc.). The organizational mode is one of balancing: one has to balance one's own, maybe conflicting needs and goals, as well as external demands. In this vein, self-management manuals encourage their readers to set their goals for themselves, to relate to them and, most importantly, to proceed methodically. However, goals become multiplied and dynamic, hence in need of constant observation and change; readers should be prepared to change their goals if circumstances, whether internal or external, require them to do so.

Setting up a relationship with oneself now means imagining oneself as the leader of an 'inner team', each member representing a specific need, capacity, or problem. The leader, that is, the self, is able to cope with both inner tensions and external demands by respectively regulating or balancing them. The relationship is thus not about *finding* oneself but about *governing* one's, possibly contradictory, and often changing, needs and demands. Identity follows a model of *corporate identity*: 'the certainty of having a strong team of various "true selves" within oneself' (Besser-Siegmund and Siegmund, 1991, p. 132 quoted in Bröckling, 2002).

In both cases, the task of working at oneself is never-ending: yet while formerly this endless task had been driven by a fundamental doubt concerning one's authenticity, it is now also driven by the view that internal and external conditions always change and therefore call for continuous monitoring and adaptive reactions.

This, albeit sketchy, *tour d'horizon* of self-help literature shows an ever-changing relationship of selves to their society, as well as a changing relationship of selves to themselves. Both kinds of relationship co-evolve: while the stratified society favored a self-relation guided by rules and obligations, postmodern society today prefers a self-regulation guided by continuously balancing heterogeneous needs and demands, both internal and external. According to the most recent variant of instruction books, governing oneself requires a truthful and trustful relation to a 'me' which is always in the making, this selfsame 'me' being the organized set of attitudes of others, which one assumes. This means hard work – *self-change work*, as it is called today. In the course of this happening, a 'me' becomes established that is capable of monitoring and regulating all kinds of demands in all dimensions of life. This is the basis – both epistemic and normative – of today's requirement to govern oneself, to be an 'operative entity'.

Yet, there is more than work: there is pleasure and passion as well. Foucault once identified the Christian roots of modern self-care in the practice of confession and its rapid spread into various domains of private and public life – he pointedly commented on this diffusion in saying that man was 'a confessing animal' (Foucault, 1990), who has come to indulge in confessing all sorts of things on all sorts of occasions in his passionate quest for his true self (see also Hahn and Schorch, Chapter 2 this volume). Not surprisingly, we find pleasure in efficient self-management too: the desire to know oneself, the aspiration to enhance oneself, or the pleasure in spoiling oneself. An overview of the most prominent techniques recommended to those who wish to govern themselves both truthfully and effectively will testify to the passion and pleasure involved. This is precisely what renders self-help such an attractive, irresistible, and uncontestable mode of governance.

Recent manuals take recourse to all kinds of techniques and carefully advise their readers to choose the most pertinent ones, depending on the problem at hand. The mode of intervention to be employed may be disciplining, permitting, discovering, or caring – or a combination thereof. Similarly, the relationship to oneself may vary, yet there is always pleasure and passion as well.

## 4   Self-help: techniques of governing oneself

How-to books instruct their readers not only how to deliberate upon the question of becoming more efficient, happier, or more successful – they also instruct their readers how to perform certain techniques, which will lead to improved efficiency, happiness, and/or success. While these techniques should ultimately become routine elements of their everyday life, the initial steps in generating a routine of, say, self-monitoring, are set up as exercises. Not unlike spiritual exercises, they require a special setting, time, and certain practices to enable diligent care for oneself. Moreover, these 'time-outs' are a matter of careful planning: frequently repeated appeals to those who consciously wish to build up zones of reflection and self-change instruct them to take time for themselves and establish a free zone reserved for themselves (e.g., Haen, 2001, p. 155).

What is more, all decisions have to be recorded in order to render them manageable and subject them to control. The most important variants of techniques are: techniques of isolation, of self-love, of knowing yourself, and of moderation. This distinction is a rather analytical one; in practice, techniques are sequenced and interact with one another: isolation will enhance self-love, and both are preparatory means of establishing a proper condition for self-inspection by moderated techniques aimed at getting to know oneself. While individual techniques can and should be chosen according to individual preferences, self-change should involve all four variants of techniques in order to be both effective and pleasurable.

*Isolation.*    The assumption is that of a fragile, highly impressionable self that is in need of focusing in order to build up a relation to him/herself. For this reason, external pressure should be reduced as much as possible. Several techniques serve this task; positive thinking gains control over your reaction to the outside world, meditation is the prevalent means of finding the way into yourself. Inside yourself, you will not find anything but silence and peace (see Osho, 2004, p. 5). On a more practical note, one should choose a room for oneself: the bathroom can be such a space, 'an oasis for self-renewal' (Pitroff, Niemann and Regelin, 2003, p. 128). 'Coddle yourself, go on an ego-trip. Everybody else gets locked out ...' (Wellfit, 2003, p. 21). This demarcation, though, is not only a spatial one. It is also a symbolic one, epitomizing one's capacity to say 'no' to the demands of others (Pitroff et al., 2003, p. 13). Cocooning practices are not necessarily anti-social, however, and a time-out can be communicated very politely: 'Five charming ways of saying No' (Pitroff, Niemann and Regelin, 2003, p. 112) show you how, for example. The pleasure you will find is relaxation and recreation – a short downtime of well-being made possible by fending off disturbing interferences.

*Self-love.*    Next to temporary isolation, in order to become independent from others, it is essential that one establishes a loving relationship with oneself. 'Be your best friend, a "nourishing mother"' (Holdau, 1999, p. 14). The bestseller, 'Your Erroneous Zone' (Dyer, 1976) offers helpful advice in the chapter entitled 'Easy-To-Master Self-Love Exercises' (Dyer, 1976, p. 44): These techniques 'included giving oneself "treats" in restaurants and stores, minimizing self-denial, joining pleasurable group activities, and standing naked before a mirror while telling oneself how attractive one was' (Dyer, 1976, quoted in Starker, 2002, p. 124). 'The loving gaze of a mother is known as transfiguring her child ... We should grant this gaze ourselves – every morning in front of the mirror. The face in the mirror will thank you with a smile' (Baur and Schmid-Bode, 2000, p. 131). 'Take, for instance, your foot fondly into your hands, provide it with security' (Holdau, 1999, p. 31). Lovingly caring for oneself is meant to create emotional independence and a trustful relation to oneself through satisfying one's longing for appreciation and comfort, security and the feeling of belonging.

*Knowing yourself.*    Techniques of introspection, reflection, and observation allow us to get to know ourselves more objectively. How-to books instruct the reader, for instance, to look at him/herself from an external point of view: 'Place yourself in front of the mirror and contemplate yourself and then respond to the following questions: Is my opposite pleasant? Do I want to be his friend?' (Tepperwein, 1997, p. 72) 'Would you employ yourself?' (Frauenpower, 2000, p. 52). In addition, how-to books encourage retrospective self-reflection: 'Each evening, before falling asleep, let the day pass before your inner eye' (Holdau, 1999, p. 19). A diary, checklists, and personality tests have come to be obligatory. Not only should the past and the present be the object of reflection, but also a vision about one's future. 'Most people lose sight of their dream of life ... retrieve it and thereby get to know yourself' (Küstenmacher, 2001, p. 290). Numerous techniques attempt to reveal one's vision and to transfer it into manageable goals and sub-goals. Regardless of whether one proceeds retrospectively or prospectively, however, the major part of such exercises relies upon writing techniques; quite literally, the issue is to commit oneself in a more objective and quasi-contractual fashion. To know yourself promises a 'delightful metamorphosis' (Küstenmacher, 2000, p. 20): enfolding the self through the improvement of self-esteem, relations to others, one's financial situation or success at work. Last but not least, an author states: 'The following seven weeks will be hard work but at the same time it will please you, because it will bring you closer to your self' (Haen, 2001, p. 61). To grant this pleasure, the same author recommends ongoing praise of the progress that occurs in self-change work. This is fully in accordance with Foucault who stated that the search for one's true self eventually becomes an object of desire and ongoing passion in itself.

*Moderation.*    Mere self-awareness does not suffice to actually cope with oneself, however. 'Working at inner stability' requires a 'domestic policy of the self' (Schmid, 2004, p. 96). A soliloquy targeted at an understanding with oneself should use traditional political means: compromise, balancing, separation of power, and the establishing of priorities (Schmid,

2004, p. 99). The prototypical technique is to imagine oneself as moderator of an inner team with the goal of balancing different needs and demands. In a quasi-democratic procedure, all partial selves have to be heard; the moderator structures the discussion and tries to find an acceptable solution. These techniques allow for inner complexity to emerge by suggesting equally complex procedures of finding an adequate solution. Instead of adamantly acting according to a strict principle, moderation is about attentively listening to all one's wishes – and hence about the pleasure of 'democratic deliberation'.

Isolation, self-love, knowing yourself, and moderation: all of these prototypical techniques enforce an ongoing process of working on oneself by addressing various aspects of the self and establishing a relationship among them. The will, the body, emotions, thoughts, and attitudes have to inform, control, or take care of each other. While the prime foci of self-work may change, the process of self-development is never to cease, but intensifies and diversifies the concept of one's self.

Isolation, self-love, knowing yourself, and moderation: all of these prototypical techniques proceed from different angles – social, emotional, cognitive, and 'political' ones. While they have to be arranged individually, depending on the problem at hand, one should not favor one type of technique and disregard others, for the techniques mutually enforce their respective impacts: in the social dimension, self-change is based on time and space for oneself; in the emotional dimension, it is about caring for oneself; at the cognitive level, it relies on knowing oneself, and at the 'political' level it is about carefully deliberating alternative options and contradictory wishes. Taken together, all techniques work at producing an incorruptible, yet trustworthy and reliable 'me' capable of enacting continuous self-monitoring and – if need be – self-change. Self-change work is highly demanding – driven forward by the promise of also being pleasurable, and it ultimately delights us with happiness, health, and/or success. It contributes to accomplishing ever more complex selves, capable of acting in an ever more complex world. This includes actively engaging with one's environment – in one's own interest and that of others (family, neighbors, community).

## 5    Becoming a neosocial self: instructions on active citizenship

Incorporating, shaping, channeling, and enhancing subjectivity, have been intrinsic to the operation of government. But while governing society has come to require governing subjectivity, this has

not been achieved through the growth of an omnipotent and omniscient central state whose agents institute a perpetual surveillance and
control over all its subjects. Rather, government of subjectivity has
taken shape through the proliferation of a complex and heterogeneous assemblage of technologies. These have acted as relays, bringing the varied ambitions of political, scientific, philanthropic, and
professional authorities into alignment with the ideals and aspirations of individuals, with the self each of us want to be.

(Rose, 1999, p. 217)

Among the assemblage of technologies, self-help practices in all their
heterogeneity are a, albeit inconspicuous yet significant, means of self-
governance. Its attraction lies precisely in the fact that although self-
helpers want to improve their self-regulating capacities, they do not
regard this practice as one of self-government that, in one way or another,
interconnects with technologies of domination. Even the seemingly
highly individual goals such as health, happiness, and success intercon-
nect with societal goals such as disease prevention and employability.
That is to say, contrary to their most individualistic appearance, all goals
pertain to individuals and collectivities alike. In this view, it will come as
no surprise that neoliberal societies deem it a matter of course that their
citizens strive for health, happiness, and success in order to become the
self-determined selves whose individual goals are compatible with societal
ones. How should this link come about? Basically, it is effectuated by
mutuality. While individual health, happiness, and success may stem
from, *inter alia*, voluntary engagement in the maintenance of their community, the society will, in turn, profit from active citizenship. Speaking
normatively, this mutuality is a matter of responsibility. The responsible
self will care for itself, as the responsible citizen will care for its community.
For in neoliberal regimes, the government of others is best linked, so it
seems, to practices in which free individuals are enjoined to govern themselves as simultaneously subjects of liberty and responsibility.

We will now, therefore, briefly examine political programs and prescriptions appealing to the individual as a free, yet responsible citizen.
In order to do so, society needs to, and indeed *can*, rely on, its citizens'
capacity to govern themselves. Active citizenship refers to the relation of
individuals and communities. 'At the core of the debate about what it
means to be an active citizen in a democracy are values, conversations and
belonging' (Taskforce on Active Citizenship, p. 11). Exerting civic responsibility, hence the promise of such responsibility, contributes to both
individual empowerment and the functionality of the community.

The need for political programs advancing active citizenship features in recent critiques on concepts of state, society, and the individual that are deemed increasingly inappropriate: as states find themselves in grave turmoil caused by globalization and monetary crises, societal cohesion seems to be endangered by increasing social inequality; modified concepts of states and societies will evolve. Although the conditions vary from country to country, a decisive feature of such new concepts is the more active role the citizens are supposed to play.

Alluding to fears and the indignation attributed to the individuals believed to be afflicted by the coldness and inconsiderateness of the present situation (Schmidt, 1998, 22), the underlying rationale is that the conception of individuals as either 'active egoists' or 'passive recipients of welfare benefits' seems to be dissatisfactory to both state and citizens. Citizens are now meant to disburden the state and promote the common good by voluntary work, oriented towards the community, and carried out free of charge. Long since a public domain, such forms of work cease to be just a supplement to state activities in terms of charity. Civic activity is rather believed to be the key to new politics that focus on 'the return of the citizen'.[2] Programmatically this was addressed in the joint statement 'Europe: The Third Way' released by British Prime Minister Tony Blair and German Chancellor Gerhard Schroeder in 1999. Whereas in Britain the 'Commission on Citizenship' addressed conceptions of this new role of the citizen back in 1988–90, it was not until 1999 that the German Parliament launched a Study Commission[3] on the *Future of Civic Activities*.

Although widely similar in terms of their outlook, the problems addressed by Blair's New Labour and Schroeder's *Neuer Mitte* seem quite contrary, yet complementary: whereas in Britain it was Thatcher's free market individualism that was to be tamed with a novel approach in tune with social democratic principles (Rose, 1999), Germany has been commonly described as ridden by a deep crisis of the social state (Sutter, 2004).

In both cases, however, the 'return of the citizens' is promoted to be the adequate remedy. Yet were citizens to return, where had they gone in the first place? Obviously, the individuals formerly supposed to be citizens of modern Western societies lack certain features incorporating the concept of citizenship now at stake. When states were active in steering all kinds of issues, and the social served as foundation for welfare, taking care of the individual and bearing risks deriving from sociality, the individual was a 'social citizen with powers and obligations deriving from membership of a collective body' (Rose and Miller, 1992: 201). Times have changed however: whereas in prospering times the state extended its tasks, it now seems overtaxed and weakened by an all-embracing will to regulate. On the one hand, limits of prosperity no longer allow these tasks to be financed, on the other hand, such state activities are conceived as practices of illegitimate paternalism. Furthermore, past experience of security and prosperity has affected the individual's commitment to political participation negatively and yielded a 'civic privatism' (Habermas quoted in German Bundestag, 2002, p. 95). This calls for the activating or enabling state. Remarkably, the term 'activating state' is used to address the support of the 'lower social third' in gaining participatory skills, whereas the model of the 'enabling state' aims at enhancing the conditions for citizens, who are already active (German Bundestag, 2002, p. 135). Modern society thus needs a state that facilitates and demands self-determined forms of action and mobilizes the acceptance of responsibility of individuals (and organizations). Put differently: it needs willing selves capable of taking decisions and taking appropriate actions. The relationship of state, society, and individual is now rearranged in such a way as to activate the individual.

This new *a priori* of political thought, however, entails two problems: the individual is supposed to look after theirself instead of expecting state benefits, and is supposed to do so because they actually *want* to do so: 'For the new politics to succeed, it must promote a go-ahead mentality and a new entrepreneurial spirit at all levels of society.' (Blair and Schroeder, 1999, p. 5)

For quite a period of time, the well-known self-management paradigm contributed to spreading an 'entrepreneurial spirit' throughout our everyday life by telling us how to take decisions and how best to realize them. It is precisely on the basis of our culturally evolved capacity to govern ourselves that neoliberal modes of governance operate. Most tellingly, a novel branch of instruction book has been issued, now occupied with how to engage in civic activities. It is important to note that these instructions rely significantly on the self-help practice: in form (*how to*), content (*self-regulation*), and goal (*empowerment*) these instructions inscribe themselves into the practice of self-change work, changing no more than the subject: this time, they are about active citizenship. In so doing, they both extend and complement the self-help genre: they add yet another domain to the project of governing oneself: next to health, happiness, and success, active citizenship may contribute to one's enhancement, if pursued methodically. At the same time, they maintain that by way of serving one's own interest, one also serves the community. A well-functioning community, in this way, is but a by-product of voluntary and individualistic action exerted by autonomous agents.

In this vein, politicians, journalists, and social scientists alike are pre-occupied with arguments as to why individuals should engage in civic activities and spread the word in almost every possible form: documents instructing municipally responsible persons, books appealing to their readers to engage actively, talk shows on the need for individual respon-sibility in times of severe political and social crisis, and so on. As diverse as the formats by which citizens are approached are the particular activ-ities they are suggested to perform, both to gain personal benefit in terms of recognition, competence building, networking, and to enhance the common good. Engagement for the young and the elderly, for nature and the community, in loose association or in fixed arrangements – the object and the particular form of engagement are highly variable in order to fit the personal preferences, resources, and interests.

As has been said earlier, the arguments in favor of such activities are hard to turn down, as they are grounded on arguments already imple-mented by the discourse on self-management: notably, they all appeal to the individual's own interest. Whereas the older motives of godliness, honor, moral conduct, or salvation are driven by altruism and duty, motives related to the self are now taking center stage:[4] gaining competences, achieving personal growth, and self-realization, to name just a few. Attending to such motives is of genuine interest to individuals: they might benefit from their activities, as they not only allow them to participate in molding social conditions, but also to create personal networks and develop individual competences. Precisely by way of promoting civic activities, virtues such as solidarity and responsibility can be established by highly acceptable and, at the same time, powerful strategies.

In Britain, for instance, the movement 'we are what we do' (2004) edited a publication that formulated the request to 'Change the world for a fiver'. No matter how incremental the advised action is, the rationale and driving *motif* is beyond contestation: all of us can help to create a different society – and what is more, in full accordance with our needs and wants. Indicatively, its subtitle pinpoints the promise of this initiative: '50 actions to change the world and make you feel good' (we are what we do, 2004).

In terms of values, the discourse on active citizenship is a prime exem-plar for the importance of focusing on 'how "values" function in various governmental rationalities, what consequences they have in forms of political argument, how they get attached to different techniques and so on' (Dean, 1999, p. 34): regularly, the authors of self-help books diag-nose times of trouble, both individually and socially. Not only are these times characterized by a lack of knowledge (knowledge society is also considered a risk society), but we also lack a universally acknowledged set of values and virtues. It is the latter, however, that the authors utilize to maintain civic activities. For, values and virtues are considered key to a good and responsible, hence meaningful, life in one's community. In this vein, Germany harbors a multitude of bestsellers dedicated to restoring freedom as the central disposition of man. By referring to various philosophical theories from Aristotle to Michael Walzer, this move is to ensure that this freedom is made use of responsibly, namely by engaging in community work (e.g., Schmidt, 1998; Wickert, 2001). In this perspective, neosocial selves need individual freedom in order to act socially.

Such a restoration is necessary because of state politics that patronize its citizens with extended measures of social provision; such is the starting point of the argument tabled by journalists, social scientists, and politicians alike. Personal freedom is the key to human existence and, frighteningly enough, endangered by state paternalism, for example, if decisions on how to arrange provision for retirement is taken out of their hands. Hence: 'By which means and measures can civic activity be promoted without undermining civic spirit, self-activity and self-will?' (German Bundestag, 2002a, p. 167). While the Bundestag deems this question to be a 'formal problem,' there is, in fact, a paradox lurking: *inducing* voluntary activities must be seen as a paradox that cannot be 'solved', but has to be processed. The report of the study commission tells us how: by readjusting formerly fixed core concepts, such as the state, the society and the citizen in such a way as to allow for a mode of ongoing action, civic education, and responsibility.

David Blunkett, then Home Secretary of the UK, refers to the Greek ideal of the city-state to argue in favor of such a *Civil Renewal*. Based on the idea of individual freedom and depending on participation and self-government, democratic institutions are meant to create and effect 'active maintenance' of this individual capacity. This capacity is to be fostered by an 'education for citizenship' and the will of the citizen to 'cultivate civic virtues' and recognize the public realm (Blunkett, 2003). The widely acclaimed demand for a return of the citizen can thus be analyzed as a 'double movement of autonomization and responsibilization' by which politics 'is to be returned to citizens themselves, in the form of individual morality and community responsibility' (Rose, 1999, p. 476). Here self-help adheres to the conception of human beings as 'ethical creatures' which is at the core of contemporary politics (Rose, 1999, p. 474).

The measures taken to bring this change about converge on what Cruikshank has termed 'technologies of citizenship', which process the paradox of undermining civic activity by promoting it. How is that achieved? Technologies of citizenship work *'through* rather than against the subjectivities of citizens' (Cruikshank, 1999, p. 69). That is, the activating state, civil society, and the good citizen are enacted by the competences, motives, and values not so much imposed upon but rather autonomously chosen by the members of neoliberal societies, for ultimately, thus the rationale, they contribute to their own enhancement and empowerment.

To be sure, personal freedom, however, is not to be taken as an absolute, such as has been claimed; the German endeavors especially characterize Thatcher's and Reagan's approaches as examples of bad practice (Schmidt, 1998, p. 81). By contrast, the use of freedom is inextricably linked to the obligation to maintain the opportunities of personal freedom, meaning to secure social order. Social order is bound to citizens acknowledging not only rights but also duties in respect of each other and the common good. Hence the rationale that as personal freedom results in the obligation to act responsibly, responsible actions indicate individual autonomy. Social cohesion then becomes the result of decisions taken by responsible citizens who act voluntarily towards the common good.

Instructions on active citizenship thus rely deeply on the culturally evolved notion of the active, self-regulated individual that autonomously, yet, responsibly, governs itself. In turn, the active self assumes yet another task: it is in the interest of its self-government to engage socially, and hence, engage in governing others – autonomously, yet responsibly.

## 6 Self-help: a socialized and sociable solitary practice in neoliberal society

'As neoliberal political rationality has spread, technologies aimed at promoting practices of the self that involve self-reflection and improvement, body monitoring and improvement, risk management, and lifestyle maximization have evolved' (Rudman, 2006, p. 186). We agree, and we hold that the governmentality perspective, inspired by the works of Foucault, proves particularly helpful to analyze the role of self-help within the framewok of neoliberal political rationality. Notably by applying a productive, relational notion of power, this perspective reveals the ways in which political power becomes effective: it operates by shaping subjectivities through albeit inconspicious, yet powerful discursive practices. These practices, however, are themselves shaped by a host of political, social and cultural factors. In this framework, self-help not only contributes to making selves capable of and responsible for governing themselves and others. At the same time, they promote the rearrangement of sociality in neoliberal society in which these normative subjectivities are center stage. They allow the state and other governmental actors to 'govern at a distance' and 'to govern through freedom' (Dean, 1994).

The governmentality perspective hence allows us to analyze in detail how this political rationality is intimately connected to and driven by specific technologies of government. These technologies exert the micopolitcics of power in that they fashion seemingly highly individual conduct, needs, and desires. In a prototypical manner, the technology called instruction books testifies to the prime *modus operandi*: they both de-scribe and pre-scribe ways of being and acting, thereby also constraining possible ways of being and acting.

At a closer look, the new modalities of governance reveal the mobilization of a new set of technologies of power, which Mitchell Dean identifies respectively as 'technologies of agency' and 'technologies of performance' (1999, p. 167). While the former refers to strategies of rendering the individual actor responsible for his or her own actions, the latter refers to the mobilization of calculative practices such as benchmarking rules that are set as parameters against which (self-)assessment can take place and which require the conduct of a particular set of performances. Small wonder, then, that governing subjectivities through liberal political choice, freedom, and autonomy comes at a price: the subject is continually confronted with norms of obligation, accountability, and responsibility. Through this process individuals now find it 'natural' to opt for a healthier, happier, more successful self and to act accordingly – because, if they don't do it (opt and act), who would or should? In this way, individuals are induced to choose active citizenship as one means of becoming an active, healthier, happier, more successful self. Induction, to be sure, operates not only by way of political programmatics promoting the positive outcomes of civic engagement. They also operate *ex negativo* by presuming citizens to be irresponsible, hence unethical, when refraining from life-long learning, prevention, and engagement – in other words, citizens who dare to possibly incur costs for society. Self-help thus becomes both an offer and an obligation for individuals – for their own sake and for that of their community.

Offer and obligation present themselves subtly and multidimensionally: when analyzing self-help manuals or instructions for active citizenship, we can see a type of language, within which objects and objectives are construed, a grammar of analyzes and prescriptions, vocabularies of programs, terms in which the legitimacy of governing self and others is established (Miller and Rose, 1993, p. 80). The discursive matrix revolves around 'goals' and 'do what you decided to do!' In this way, self-help books are revealed as one particular practice by which subjects regulate themselves and others – they represent, in fact, a governmental technology of producing neosocial selves.

Key to maintaining a social self is discipline and control. The self must be managed through constant self-monitoring and self-regulation to ensure that it does not stray from the ideal. As the ideal is unattainable, so the process is a constant battle with the self to reach that elusive goal. It is driven on by the appeals to and imagery of being an 'autonomous chooser' in a 'freely operating community'. If anything should constrain the autonomous chooser, it is these communities that are thought to provide them with more specific guidelines as how to, in fact, enact their freedom. Next to responsibility the binding force is neighboring values such as shame, guilt, obligation, or duty. Nikolas Rose terms these new politics of behavior 'ethopolitics'. Ethopower 'works through the values, beliefs, and sentiments thought to underpin the techniques of responsible self-government and the management of one's obligation to others' (Rose, 2000, p. 1399). Following Foucault, Rose discerns two broad configurations of such politics. 'The first seeks to govern the ethical self-regulation of the individual in terms of fixed moral codes justified by relation to some external set of principles or concepts of human nature. The second emphasizes the aesthetic elements in the government of ethics; the self-crafting of one's existence according to a certain art of living, whether this concerns friendship, domesticity, erotics, or work. In contemporary ethopolitical debate, one can see elements of each of these configurations.' (Rose, 2000, p. 1399)

It is particularly the theme of 'responsibilizing the self,' a process at once economic, political, and moral, which is emblematic to the ethopolitical shift in the governance of neoliberal society. Notably, 'the duty to the self – its simultaneous responsibilisation as a moral agent and its construction as a calculative rational choice actor – becomes the basis for a series of investment decisions concerning one's health, education, security, employability, . . .' (Peters, 2001, p. 43). *Voilà* the enterprising selves in an 'enterprise culture' (Keat and Abercrombie, 1991).

'To a large extent, liberal modes of governance in enterprise culture render the lives of individuals as private matters free from state intervention by offering citizens the opportunity for "choice", for "autonomous" life plans and the "freedom" to be the persons they want to be' (Rimke, 2000, p. 72). This produces a twofold result: on the one hand, citizens are being addressed as being free to and capable of making choices as to their personality and entire life plans. On the other hand, however, the procedures to bring these choices about are deeply intertwined with strategies and obligations of neoliberal governance: both effects are connected by appeals to individual responsibility.

Ultimately, this accounts for the fact that in current society, its neoliberal rationality notwithstanding, the concept of the social has not dissolved: by contrast, it is now realized by way of responsible citizens. How? The practices and technologies connected to the autonomous chooser, which have come to permeate our society, are both *socialized* and *sociable* techniques of governing oneself and others. As self-help equips us with this conception of ourselves and others it produces accountable 'addresses'. Such addresses dispose of a self-referential account encompassing the notion of individual responsibility (see Nassehi, Chapter 4 in this volume). By so doing, the genre allows for the mutuality of expectations and the expectation of expectations: while they grant structural flexibility, they simultaneously secure operative stability. The result may thus turn out to be highly individualistic, the procedure, however, is not.

In this respect, the genre as a whole proves part and parcel of societal differentiation, rather than being dismissed as idiosyncratic outbreaks into excesses of self-care (which, in some cases, may be true). It ultimately contributes to producing a culture of individuals who, *vis-à-vis* highly specific choices as to their conduct in life, still manage to remain sociable – just by way of adhering to highly conventionalized procedures of self-care. Indeed, our culture sees itself as a culture of individuals, yet according to Niklas Luhmann, this implies that individuals have to discipline themselves accordingly. If social order and reciprocal expectations are to remain possible (Luhmann, 1992, p. 199), self-discipline is indispensable. What renders self-discipline acceptable, however, is its current form: neosociality is based upon the increasingly accepted technologies realizing neoliberal rationality.

Indeed, the remarkable rise and spread of neoliberal political rationality has rapidly come to shape ourselves, our society, and the relation between both: more specifically, it is already on the verge of becoming self-evident, if not 'common sensical', and is, therefore, particularly powerful. High time to scrutinize its ruling principle – freedom – for what it is: a *technology*:

> Under neoliberal conditions freedom becomes a technology of freedom ... this means that freedom once more is a matter of networks of freedom that are integrated with our existence. This is, of course, no

absolute freedom – whatever this could be – but we talk about networks of trust, of risk, of choice. Networks that invite us to overcome the incalculability of our lives by way of entrepreneurship and acts of free will ... (on this understanding) freedom has a price: continuous monitoring. Wherever freedom appears in our neoliberal era, there is monitoring, audit, regulation of norms. In other words: *forms of freedom that integrate us with the whole continuum of acceptable forms of (self-) government.*

(Osborne, 2001, p. 15; our emphasis, SM, BS, and SD)

Self-help as a sociable and socialized practice forms part of today's accepted forms of (self-) government. Although a passage from disciplinary societies to societies of control (Deleuze, 1990) can be observed, mechanisms of control do not dispose of disciplinary mechanisms. Disciplinary deployments coexist with regimes of control. Most decisively, the society of control establishes an audit culture (Power, 1997), in which a 'calculative regime' prevails (Miller, 1992). Willing selves, capable of governing themselves and/with others, are embedded in this very regime, which seizes upon, polices, and controls this capacity – in most helpful ways.

## Notes

1. Sections 3 and 4 are based upon Maasen and Duttweiler (forthcoming).
2. Kymlicka and Norman (1994) use this phrase to describe recent developments in both political agenda-setting and discussions of political philosophy.
3. It refers to the British efforts but also to similar initiatives in the Netherlands and the USA (German Bundestag, 2002, p. 42).
4. Indeed, self-centered motives are the temptation, despite which honorary work is to be performed.

## References

Baur, G. and W. Schmid-Bode, *Glück ist kein Zufall. Lassen Sie sich vom Glück berühren. Die besten Methoden für ein erfülltes Leben* (München: Gräfe und Unzer, 2000).

Becker, R. Z., *Not- und Hilfsbuechlein fuer Bauersleute* (Leipzig: Göschen, 1787).

Blair, T. and G. Schroeder, *Europe: The Third Way* (1999). Accessed 22 December 2006 at: http://www.socialdemocrats.org/blairandschroeder6-8-99.html

Blunkett, D., *Civil Renewal: A New Agenda* (London: Home Office Communication directorate, 2003), accessed July 2006 at:
http://www.active-citizen.org.uk/files/downloads/active_citizenship/EKML.pdf

Bröckling, U., 'Das unternehmerische Selbst und seine Geschlechter. Gender-Konstruktionen in Erfolgsratgebern', *Leviathan*, 48, 2 (2002): 175–94.

Carnegie, D., *How to Win Friends and Influence People* (Kingswood: The World's Work, 1938/1977).

Cruikshank, B., The *Will to Empower: Democratic Citizens and Other Subjects* (Ithaca, NY, and London: Cornell University Press, 1999).

Dean, M., *Critical and Effective Histories: Foucault's Methods and Historical Sociology* (London and New York: Routledge, 1994).

Dean, M., *Governementality: Power and Rule in Modern Society* (London: Sage, 1999).

Deleuze, G., 'Society of Control', *L'autre journal*, I (1990), accessed Janurary 2007 at: (www.nadir.org/nadir/archiv/netzkritik/societyofcontrol.html

Duttweiler, S., 'Professionalisierung von Orientierungswissen? – Lebenshilferatgeber als Experten der Lebensführung', in K.-S. Rehberg (ed.), *Soziale Ungleichheit, kulturelle Unterschiede. Verhandlungsband des 32. Kongress für Soziologie* (Frankfurt am Main and New York: Campus, 2006), pp. 3192–201.

Duttweiler, S., *Sein Glück machen. Arbeit am Glück als neoliberale Regierungstechnologie* (Konstanz: UVK, forthcoming).

Dyer, W., *Your Erroneous Zones* (New York: Funk Wagmalls, 1976).

Foucault, M., 'Technologies of the Self", in L. H. Martin, H. Gutman and P. H. Hutton (eds), *Technologies of the Self: A Seminar with Michel Foucault* (London: Tavistock, 1988), pp. 16–49.

Foucault, M., *The History of Sexuality: An Introduction* (New York: Vintage Books, 1990).

Foucault, M., *The Politics of Truth* (Cambridge, MA: Semiotext(e), 1997).

*Frauenpower für Powerfrauen. Der beste Weg zu mehr Glück, Erfolg und Selbstbewusstsein* (München: Südwest Verlag, 2000).

Gerling, R., *Die Gymnastik des Willens. Praktische Anleitung zur Erhaltung der Energie und Selbstbeherrschung, Kräftigung von Gedächtnis und Arbeitslust durch Stärkung der Willenskraft ohne fremde Hilfe* (Berlin: Moeller, 1920).

German Bundestag, *Bericht der Enquete-Kommission 'Zukunft des Bürgerschaftlichen Engagements'. Bürgerschaftliches Engagement: auf dem Weg in eine zukunftsfähige Bürgergesellschaft* (Opladen: Leske and Budrich, 2002).

Giddens, A., *Modernity and Self-Identity: Self and Society in the Late Modern Age* (Cambridge: Polity Press, 1991).

Giddens, A. *The Transformation of Intimacy: Sexuality, Love and Eroticism in Modern Societies* (Cambridge: Polity Press, 1992).

Greenberg, G., *The Self on the Shelf. Recovery Books and Good Life* (New York: State University, 1994).

Grossmann, G., *Sich selber rationalisieren. Mit Mindestaufwand persönliche Bestleistungen erzeugen* (Stuttgart: Verlag für Wirtschaft und Verkehr, 1927).

Haen, R., *Das Zicken-Prinzip. Der weibliche Weg zu Ruhm und Glück* (München: Econ Ullstein List, 2001).

Helmel, H., *Kraft und Tat. Wegweiser zu Wille, Gesundheit, Kraft* (Passau: Wegbereiter-Verlag, 1928).

Holdau, F. Einfach gut drauf. Tolle Gute-Laune-Macher. Wellness-Tips und Psycho-Tricks. (München: Gräfe und Unzer, 1999).

http://www.phoenixcentre.com/articles/practical_intimacy.htm

Knigge, A., v. *Über den Umgang mit Menschen* (Frankfurt am Main: Insel Verlag, 1788/1977).

Koch-Linde, B., *Amerikanische Tagträume. Success und Self-Help Literatur der USA* (Frankfurt am Main and New York: Campus, 1984).

Küstenmacher, W. T., *Simplify your Life. Einfacher und glücklicher leben* (Frankfurt am Main: New York: Campus, 2001).

Kymlicka, W. and W. Norman, 'Return of the Citizen: A Survey of Recent Work on Citizenship Theory', *Ethics*, 104, 2 (1994): 352–81.

Lessenich, S., 'Soziale Subjektivität. Die neue Regierung der Gesellschaft', *Mittelweg 36*, 4 (2003): 80–93.

Lindworsky, J., *Willensschule* (Paderborn: Schöningh, 1927).

Luhmann, N., *Beobachtungen der Moderne* (Opladen: Westdeutscher Verlag, 1992).

Maasen, S., *Genealogie der Unmoral. Zur Therapeutisierung sexueller Selbste* (Frankfurt am Main: Suhrkamp, 1998).

Maasen, S. and S. Duttweiler, 'Intimacy in How-To Books and the Passion of Self-Change Work', in E.L. Wyss (ed.), *Transformations of Passion: The Mediatisation of Intimacy in the 20th Century* (Amsterdam and New York: John Benjamins, forthcoming).

Miller, P., 'Accounting and objectivity: The invention of calculating selves and calculable spaces', *Annals of Scholarship*, 9 (1992): 61–86.

Miller, P. and N. Rose, 'Governing Economic Life', in M. Gane and T. Johnson (eds), *Foucault's New Domains* (London and New York: Routledge, 1993), pp. 75–105.

Moskowitz, E. S., *In Therapy We Trust: America's Obsession with Self-Fulfillment* (Baltimore: Johns Hopkins University Press, 2001).

Newman, M. and B. Berkowitz, *How to Be Your Own Best Friend* (New York: Ballantine, 1971).

O'Malley, P., 'Risk and Responsibility', in A. Barry, Th. Osborne and N. Rose (eds), *Foucault and Political Reason: Liberalism, Neo-liberalism and Rationalities of Government* (Chicago: The University of Chicago Press, 1996), pp. 189–207.

Osborne, Th., 'Techniken und Subjekte: Von den "governmentality studies" zu den "studies of governmentality"', *IWK-Mitteilungen*, 2–3 (2001): 12–16.

Osho, *Zeitschrift des Osho-Institut Köln* (Koln: Osho-Institut, 2004)

Peters, M., 'Education, Enterprise Culture and the Entrepreneurial Self: A Foucauldian Perspective', *Journal of Educational Enquiry*, 2, 2 (2001): 58–71.

Pitroff, U., C. Niemann and P. Regelin, *Wellness. Die besten Ideen und Rezepte für die Wohlfühloase zu Hause* (München: Gräfe und Unzer, 2003).

Pittrof, Th., *Knigges Aufklärung über den Umgang mit Menschen* (München: Wilhelm Fink, 1989).

Power, M., *The Audit Society: Rituals of Verification* (Oxford: Oxford University Press, 1997).

Riesman, D., in collaboration with R. Denney and N. Glazer, *The Lonely Crowd. A Study of the Changing American Character* (New Haven: Yale University Press, 1950).

Rimke, H. M., 'Governing Citizens Through Self-Help Literature', *Cultural Studies*, 14, 1 (2000): 61–78.

Rose, N., *Governing the Soul: The Shaping of the Private Self* (London: Free Association Books, 1999).

Rose, N., *Powers of Freedom: Reframing Political Thought* (Cambridge: University of Cambridge, 1999).

Rose, N., 'Governing the Enterprising Self', in P. Heelas and P. Morris (eds), *The Values of the Enterprise Culture: The Moral Debate* (London and New York: Routledge, 1992), pp. 141–64.

Rose, N., 'Inventiveness in Politics', *Economy and Society*, 28, 3 (1999): 467–493.

Rose, N., 'Community, Citizenship, and the Third Way', *American Behavioral Scientist*, 43, 9 (2000): 1395–1411.

Rose, N. and P. Miller, 'Political Power beyond the State: Problematics of Government', *The British Journal of Sociology*, 43, 2 (1992): 173–205.

Rudman, D.L., 'Shaping the active, autonomous and responsible modern retiree: an analysis of discursive technologies and their links with neo-liberal political rationality', Ageing & Society, 26 (2006): 181–201.

Sartorius, J., *Die Feldherrnkunst des Lebens. Eine Willensschule* (Paderborn: Schöningh, 1929).

Schmid, W., *Mit sich selbst befreundet sein. Von der Lebenskunst im Umgang mit sich selbst* (Frankfurt/M.: Suhrkamp, 2004).

Schmidt, H., *Auf der Suche nach der öffentlichen Moral. Deutschland vor dem neuen Jahrhundert* (München: DVA, 1998).

Sprenger, R., *Die Entscheidung liegt bei Dir!* (Frankfurt/M.: Campus, 1997).

Starker, S., *Oracle at the Supermarket: The American Preoccupation with Self-Help Books*. (New Brunswick: Transaction, 2002).

Sutter, B., 'Governing by Civil Society: Citizenship within a New Social Contract', in J. Angermüller, D. Wiemann, R. Kollmorgen and J. Meyer (eds), *Reflexive Representations: Politics, Hegemony, and Discourse in Global Capitalism* (Münster: LIT, 2004), pp. 155–68.

Taskforce on Active Citizenship, *Background Working Paper* (Dublin), accessed 22 December 2006 at: http://www.activecitizen.ie/index.asp?docID=61.

Taylor, C., *Sources of the Self: The Making of the Modern Identity*, (Cambridge: Harvard University Press, 1989).

Taylor, C., *The Ethics of Authenticity* (Cambridge: Harvard University Press, 1991).

Tepperwein, K., *Der Schlüssel zum Glück. Ein Anti-Ärger-Programm* (Güllesheim: Die Silberschnur, 1997).

Trilling, L., *Sincerity and Authenticity* (London: Oxford University Press, 1982).

We are what we do, *Change the World for a Fiver: 50 actions to change the world and make you feel good* (London: short books, 2004).

*Wellfit*, 2 (2003).

Wickert, U., *Zeit zu handeln. Den Werten einen Wert geben* (Hamburg: Econ, 2001).

# 2
# Technologies of the Will and Their Christian Roots

*Alois Hahn and Marén Schorch*

The question of free will and personal responsibility has become more important in recent times, especially encouraged by the theories of neurologists like Wolf Singer and Gerhard Roth (2003, p. 33).[1] Their theses can be summarized as follows: freedom and responsibility are all hollow words, pure figments. This is not because they depend on circumstances but rather on our brain – is this the return of the brain-mythology of the nineteenth century? Singer and Roth are not only in academic debate but are also considering the practical consequences for everyday life from their empirical results. Their deterministic assumption, however, would not affect social practice, as social actions are judged and understood by their consequences rather than by their anteceding factors.[2] In social practice, the latter would amount to nothing less than 'predestination'. Interestingly enough, to believe in the predestination of our actions and the feelings of our brains would be quite comparable with the traditional belief in predestination by 'God's brain', so to speak 'cerebrum cerebrorum'. Of course, the question of the freedom of our will and the resulting problem of responsibility for our actions and failures is much older than brain research. It has accompanied European philosophy and theology since their beginning and was of central importance for the emerging of sociology. Sociologists as well as philosophers and theologians have dealt with these questions (not only, but certainly in the context of Christianity on which we are focusing here), even if they have not been involved in the actual debate about brain research. However, we do not intend to enter this discussion in our chapter; we relinquish brain research to the experts of this field of research. Nonetheless, a sociological perspective *on* the debate could perhaps offer a 'second-order observation' (a more complex view of the influences between neurophysiological research and society, and *vice versa*).[3]

Even though the problem of responsibility is connected to the problem of free will, they are not similar in every detail. Scholars of different periods agonized over this question, and Erasmus, for example, mentioned this old and spiny question in a text about 'Liberum arbitrium' targeted at Luther,[4] 'Inter difficultates, quae non paucae occurrunt in divinis literis, vix ullus labyrinthus inexplicabilior quam de libero arbitrio. Nam haec materia iam olim philosophorum, deinde theologorum etiam, tum veterum, tum recentium ingenia mirum in modum exercuit, sed maiore, sicut opinor, negotio quam fructu.' Experiencing actions, associations, and motives as highly contingent to the social environment as well as to the agent is a central presumption. Others are as surprising to us as we are to them. As we never lose the feeling of contingency, we are as surprising to ourselves as well, as we and others stay undisclosed (see Hahn, 1997 and 2002):

> Since one never can absolutely know another, as this would mean knowledge of every particular thought and feeling; since we must rather form a conception of a personal unity out of the fragments of another person in which alone he is accessible to us, the unity so formed necessarily depends upon that portion of the Other which our standpoint toward him permits us to see. These differences, however, by no means spring merely from differences in the quantity of the apprehension. No psychological knowledge is a mere mechanical echo of its object. It is rather, like knowledge of external nature, dependent upon the forms that the knowing mind brings to it, and in which it takes up the data.
> (Simmel, 1906, pp. 442f.)

Ascriptions of responsibility to God or social setting, brain or devil, fate or restraints of authority or economy are interpreting those experiences of contingency in completely different ways, but they are always within the context of determination of the undetermined. Niklas Luhmann assigned this function to religion (see Hahn, 2001). What does that mean for free will and responsibility in the context of Christianity? Dealing with the question of individual actions and social structures, sociology is preoccupied with personal identity, with the self, acquired in social practices. We will investigate this aspect in greater detail in the following sections, focusing on one of the oldest social practices in the religious context – confession, as biography generator. Our thesis is that changing structures of society presuppose different forms of social control and various forms of biographical identities for the constituent populations, and that the history of institutions obliging people to confess can be considered as evidence of this general hypothesis. Starting with general remarks about personal identity and self-thematization (Section 1) as well as confession of acts (Section 2), our main part will systematically analyze the dimensions of confession as technology of the will, such as defining sins against God's will, willful inspection of the sinner, and inducing her/him to confess sins (Sections 3–6). Finally, we will discuss the importance of confession as biography generator in modernity (Section 7).

## Self and self-thematization

First, we have to elucidate what a certain thematization or responsibility refers to: the self (or personal identity,[5] to use a more sociological term). Self-conception of a person in reference to its own individuality and polarization covers 'positive marks or identity pegs, and the unique combination of life history items that comes to be attached to the individual with the help of these pegs for his identity' (Goffman, 1963, p. 74). Using those 'identity marks/pegs' such as name, certification of birth, fingerprints, or handwriting allows us to ascribe a biography to a certain person in a certain situation, to identify someone as unique.[6]

According to social behaviorism and its prominent founder, George Herbert Mead, identity or self is not naturally given, but acquired in socialization and thereby socially constructed (Mead, 1934). At the center of his analysis, we find the differentiation between *I* and *me* that, in combination, constitute the *self* of an individual. To specify this: *I* corresponds to the biological nature of man (the spontaneous reactions as well as drive equipment) whereas *me* is the organized set of attitudes of others which one assumes. These attitudes of others constitute the organized 'me', whereas reactions towards it are defined by the *I*. Mead strictly emphasizes the importance of the group surrounding the individual: 'Now, in so far as the individual arouses in himself the attitudes of the others, there arises an organized group of responses. And it is due to the individual's ability to take the attitudes of these others in so far as they can be organized that he gets self-consciousness. The taking of all those organized sets of attitudes gives him his 'me'; that is the self he is aware of' (Mead, 1934, p. 175).

In addition to Mead's concept of personality, the self can be divided into an *implicit* and an *explicit* form too:[7] the implicit self covers all acquired habits, dispositions, experiences and so on that are characteristic and formative for the individual – the self as 'habitus ensemble' (Hahn, 2000b, p. 99). This facet of the self is historically universal and not reflexive to the self in a narrow sense, quite in contrast to the explicit self that – nomen est omen – elevates his person explicitly in description and communication.

Here, the self emerges as a result of communicative constructions. This self-thematization cannot only be observed in diaries or interviews (we come back to this later), but also in everyday life when a person refers to a self in a certain situation that endures beyond it or when a statement is defined as 'typical' (or untypical) for this person. So we speak of 'situational self-thematization' by the strict attachment to a particular situation, whereas 'biographical self-reflection' refers beyond and is not at all universal. How exactly someone manages to relate to themselves as object depends heavily on the existence of institutionalized forms of self-thematization in a society – 'biography generators' – as suggested in earlier works (Hahn, 1987). As already mentioned, confession is an important example for those institutions (apart from diaries, psychoanalyses, autobiographies, and different types of qualitative interviewing methods, especially biographical in-depth interviews) and will be at the center of our following analysis.

Biography generators are socially institutionalized devices that generate special kinds of discourse, in this case confessions, where the main topic is the biography of at least one of the participants. The theme of these discourses is, so to speak, a narrative form of identity or, in other words, the self as a tale told. It cannot be taken for granted that people speak about themselves, of their lives, emotions, intentions, acts, and so on, nor can it be taken for granted that they find an audience for such stories. The kind of biographical identity or textual self one possesses depends on the historically and culturally varying forms of biography generators. If these generators are almost totally lacking, the self cannot take the form of a narrative identity. This kind of identity presupposes, in other words, institutions of self-thematization or a social 'memory function'.

Naturally, people everywhere have always had a life course, a 'Lebenslauf,'[8] but this does not necessarily mean that they have a biography, which always constitutes a narrative selection of relevant facts or fantasies and may start earlier or end later than the 'real' biological life span of the concerned person (the famous Italian artist Cellini starts his biography with the birth of Julius Caesar, and Goethe has his begin with the description of the constellation of the stars).

The identity-generating function of memories is not without paradoxical effects: first, since it mainly presents the past, or at least a part of it, as the essential present – what is really present is no longer present. Second, while a biography extends over time, the narration has to deal with the sequentially disjointed, as if it were a simultaneous totality. One might call this the 'Tristram Shandy Paradox'. In all reflexive identification, the self functions are at the same time subject *and* object of a discourse. The whole has to be presented by a part of the whole. At any rate, the self resulting from biography generators is a reflexive entity and thus, inevitably, a paradox.

Biographies can be read as a manifestation of the free will of their 'owner,' as the unintentional consequence of fate, or, last but not least, as the manifestation of predestination. The auricular confession, of course, as conceived by Catholic authors since the Middle Ages (especially since Abelard), emphasized relatively free will as essential for one's deeds, whereas the Protestant tradition since Luther, and especially in the Calvinist and Puritan context, stressed the role of grace. Whatever theory was accepted in these religious traditions, responsibility for one's life could never be totally denied.

One of the earliest discourse generators in Christian societies was, as we see it, the auricular confession in the Christian church that obliged the general population to speak about themselves. We will now take a closer look at historically changing forms of this institution. As mentioned in the introduction, we argue that changes in the structures of society require different forms of social control and of biographical identities for the constituent populations, and that the history of institutions obliging people to confess can be considered as an example of this general hypothesis. One reason for the importance of the confession of sins as a universal biography generator is, of course, its ability to be generalized.

## The confession of acts

In the age of early Christianity, when the self-definition of the Christian church was that of a community of saints or religious 'virtuosi' in the Weberian sense, there was virtually no need for confession at all. Ideally, such a community could not accept that its members perpetrated deadly sins. Such self-definition could not be maintained when the church became a mass organization. Sinning had to be reckoned with as an everyday affair, because membership could no longer be reserved for the elite of allegedly pure saints. The church had become a community comprising a majority of ordinary sinners. The institutional form of social control applicable to this situation was the invention of the auricular confession to the priest. From the fifth to the twelfth centuries what had to be confessed were deeds; the motives did not play such an important role. The penitentials, the handbooks guiding the father confessors, contain lists of acts and corresponding punishments and restitutions, but did not try to find out what the relevant motives were. The rank-and-file member was not asked about the *inner state of his soul*. This identity generating privilege, although questionable, was reserved for the clerical elites. The identity of the common individual as the result of narrative self-disclosure in institutionalized contexts did not encompass motives or constitute a biography as a temporalized sequence of identity-constituting events. It seems due to this that the free will is not of any relevance here.

The situation changed drastically in the twelfth century. The material condition of society that had been rapidly evolving during this era, as evident in the rise of cities, increased local mobility, supra-regionality of trade, stronger differentiation of professions, the appearance of important opportunities for individual initiatives, the evolution of literacy and institutions of higher learning leading to the foundation of the universities, the first stages of a supra-local market capitalism, and, last but not least, the foundation of the supra-regionally organized mendicant orders. The old confessional form of social control and understanding could no longer be restricted to exterior deeds. The most influential intellectual reaction to these changes was a third way of looking at guilt and confession.

## The confession of motives

Central to this is Abelard's theory of sin, which held that sin is not really tied to external action; rather its core lies in an act of intention and will, in the agreeing to sin. It is only such agreement that leads to guilt of the soul, and it is this that deserves damnation, because one has become guilty before God (see Abelardus, 1836, p. 211). The radical relocation of sin within man is in the strongest contrast to earlier conceptions in which a more external definition of guilt was the norm. As Jacques Le Goff so precisely describes, the world of the Middle Ages is an extroverted world. External duties and failures are at the center of ethical attention (see Le Goff 1977, p. 167). The Church's view of sin fits this opinion: there is a correspondence between sins viewed as external acts and repentance oriented toward external retaliation: confession in church is a confession that involves a scale, since it defines the punishment, without taking into account the motives, relative to the severity of the act. The focus of early medieval church confession is therefore not the confession as such, but rather the reparation (*satisfactio*) that follows the confession.[9]

The twelfth century's new definition of sin necessitated a corresponding internalized form of repentance. The sinner gains real forgiveness by eliminating the inner reality of the sin through negation of intention, that is, through deep remorse. This remorse (the technical term is *contritio*) is based neither on external factors nor on the fear of eternal or finite punishment, but rather on the realization of the shamefulness of the sin. It is based on the pain felt over having had such intentions. Experiencing regret out of the love of God is itself a gift of God; it eliminates the guilt, and with it, the punishment. The repentant soul no longer requires punishment (see Abelardus, 1836, p. 628).

Nonetheless, the sinner is encouraged to confess his guilt. Confession does not become superfluous or irrelevant through the internalization of guilt; it rather now becomes a forum to which not only external acts but also intentions are brought. This provides a form of social control of feelings and of conscience which was not possible before. Thus, church confession had become an institution before which the individual must justify himself. The shift of guilt to the realm of intentions soon becomes (with a few modifications that are not relevant here) a common component in scholastic philosophy. More than ever before, the individual is thrown back to introspection to the degree that the ideas in question became predominant. The individual's *inner motives* are tied to his salvation and therefore need to be explored (hence we can speak of confession by will). Illumination of one's own web of motivations is connected to an intensification of the sense of one's own subjectivity that provides a new way of becoming aware of one's self-history.

## Institutionalization of mandatory church confession

These new contents of consciousness did not remain mere intellectual responses to a new situation; they became elements of the reality of institutions. The Fourth Lateran Council of 1215 emphasized the duty of every Christian, independent of their sex, to confess before the local priest at least once a year. This amounted to the founding of an institution in which theories became effective in practice in terms of pastoral duties. Previously these theories were only discussed or accepted in theologians' circles and universities. Now, the new teachings of guilt and responsibility began to gain influence as an instrument of discipline and a tool that established meaning. One could not avoid confession easily. The Church was the institution that had the monopoly for granting access to salvation, and, as such, it firmly established confession.

The already mentioned increase in individualization can be witnessed in many documents in the twelfth century, and more explicitly in the thirteenth and fourteenth centuries. This was not the direct result (as far as it can be explained by the variables chosen in this chapter) as a new theory of guilt, but rather of its translation into a societal institution that is not primarily defined by its coercive character but is nonetheless unavoidable. Certainly there were efforts even before 1215 to encourage believers to receive the sacraments regularly. As a case in point, consider that in 506 the regional council of Agde required receiving the Last Supper at least three times a year. However, decrees of this kind remained ineffective. In addition, theologians continued to debate whether oral confession was necessary after repentance. The canon *omnis utriusque,* however, was strictly enforced; Peter Browe provides evidence of this (Browe, 1933). In regions of France with a high concentration of heretics, the failure to receive the sacraments was taken as indication of membership in a heretical group, because they rejected the monopoly of priests over the administration of sacraments. The rejection of the sacraments also led to the institution of the Inquisition.

When mandatory confessions were made outside the home district, the confessor had to submit proof and often had to obtain permission from his or her local priest. After the Middle Ages it became customary to give a receipt to each confessor. In many cases the receipt also had to be presented to the secular authorities, as was the case in Austria during the seventeenth and eighteenth centuries. Delinquency was punished with fines or by being refused alms. But it was not only external pressure that, during the Middle Ages, led to the mass institutionalization of regular church confession. It is probably at least as important that the conviction gradually became widespread that the sacraments concerning penitence were necessary for salvation and that they were a gift of God.[10] In particular, the belief grew that all sins (certainly all deadly sins) must be confessed. The intentional concealment of a deadly sin invalidated the entire confession and added another deadly sin to those already committed. If a deadly sin was unintentionally forgotten, it had to be confessed later. Not only that, but, according to some theologians, in addition to the deadly sin having to be confessed, the entire confession had to be repeated. This was necessary at least in those cases where the second confession was not performed by the same priest, or when he failed to recall the first confession.[11]

As the mandatory confession spread, the theory of repentance was modified. In Abelard only the total repentance (*contritio*) played a role, but as from the thirteenth century the idea of incomplete repentance (*attritio*) emerged. This was repentance out of fear. The sinner imagines that the punishment for his deeds is waiting for him in the 'beyond'. Increasingly, the fear of purgatory and hell came to be interpreted as sufficient repentance for sins, if the fear was created in the context of a confession. With this, repentance was no longer recognized solely in the form of which only a virtuoso of religion was capable; it was connected to motivations that had meaning for the normal lay person as well. In addition, as incomplete repentance became more institutionalized, fear was used in those cases as a regulatory mechanism. Earlier, Thomas Aquinas was skeptical about the effectiveness of mere *attritio* for achieving salvation. However, Duns Scotus was of the opinion that a minimum of *attritio* is a sufficient condition for the effective reception of the sacraments. The definitive institutionalization of the role of *attritio* occurred during the Council of Trent.[12]

One of the most concrete results of the introspection that was required by the duty to confess was a new sensitivity for the uniqueness of the individual. This uniqueness has been expressed in the individualization of sculptured headstones since the thirteenth and fourteenth centuries. Philippe Ariès shows that this enhanced individualization of identity consciousness was also partly responsible for a changed view of death that became ever more pronounced from the twelfth century. Earlier periods had 'mastered' death and had experienced it mostly as a group experience; now death was beginning to be experienced as an individual crisis. The archaic control of the fear of death did not need to be as strong because the individual knew himself to be an integral part of the surviving group that continues to live. However, when death destroys an individual who sees himself as a unique person, the fear of death must become much more dramatic. Ariès points out that this new understanding of death, 'la mort de soi,' was responsible for a new conceptualization of the belief in a 'beyond'. In particular, there was general acceptance of the view, which used to exist only among the theological elite, that body and soul will be separated immediately after death, the conviction of an individual reckoning at the moment of death, and a general belief in purgatory (see Le Goff, 1981).

A further important corollary concerned the existence and designation of rules that guide the exploration of one's own conscience. The individual would not know where to begin his inner exploration, were he not given a map of the landscape of his soul. It is of great importance for the development of confession according to the Fourth Lateran Council, that a wealth of manuals to be used by the priest was prepared. These casuistically surveyed and systematized the world of sin, virtue, intentions, motives, and the degrees of freedom and responsibility. Most of these texts may be subsumed under a literary category: the so-called 'Sums' for priests, the 'Summae confessorum' or 'Summae de Casibus conscientiae'.[13] Raymundus of Penaforte is generally recognized as the founder of this genre. His Summa (the so-called 'Raymundina') was written in the decades following the Lateran Council. Some of these manuals, showing clear differences from one author to another and from one era to another, remained in use for centuries. Differences may appear in the evaluation of the severity of certain sins, the intensity with which the circumstances surrounding the act are dissected, and the web of motives. Despite these differences, however, a high degree of similarity remains. The main function of the manuals was to provide the priest and confessor with moral certitude for the evaluation of the ethicality of acts and motivations. In times when action and behavior were becoming more and more complex and differentiated, increasingly detailed specifications of general moral principles represented templates for interpretation that allowed the individual to orient himself *vis-à-vis* the new wealth of possibilities and to overcome the anxieties resulting from guilt. Like clients on the psychotherapist's couch, confessors were able to find a map for the evaluation of their sins in the casuistry of the Sums, manuals, and rule books for confession. The priest was instructed to inquire about sins in general, and also to bear in mind that, typically, the prince has to combat sins that differ from those of the knight, the merchant, the burgher, and so on.

## Biographical confession

The fourth phase of the development of confession was reached with its institutionalization during the Reformation and the Counter-Reformation. One of the most consequential changes brought about by the reformation of religious life in Europe is probably the redesigning of the confession. Generally speaking, the character of confession as a sacrament was abandoned. Also abandoned was the idea of the priest's magical powers to forgive the repentant sinner's guilt. As soon as salvation became based on belief or predestination, regular confession lost its relevance as cleanser of sins. But that did not necessarily mean that it lost its relevance totally. In any case, the Reformers did not eliminate confession; what they did change were its forms, its theological significance, and ethical function. In particular, two elements became autonomous: on the one hand, the searching of one's conscience, that is, the individual's examination of his own beliefs and worthiness to be forgiven and, on the other hand, the surveillance of the external that was practiced by the priest or the community.

Luther's attitude towards confession developed gradually, and toward the end of 1521 he arrived at his final view of it. Thereafter, the sacramental character of confession and its necessity for salvation became an issue of debate. Oral confession is not documented in the Bible; therefore it is a human creation. Nonetheless, Luther was not in favor of abolishing it. Fischer summarizes Luther's view as follows:

> Despite all this, institutionalized confession remains a very wholesome institution, and every Christian will make use of it voluntarily and gratefully. However, there cannot be coercion in any way. It is entirely

illegitimate to require a confession that goes into the smallest detail and has to be given before the priest ... Confession must be put under the will of the individual; in particular, the confession before a lay-person must explicitly be allowed ... what is required is only the confession before God. He who performs it in the right manner will also feel the need to confess to a brother in Christianity and will receive many blessings from doing so. But he should confess to the priest as only a brother-in-Christianity and not as someone who holds public office and has special privileges ... In all this, the emphasis must be on the right attitude of the heart, and on the right and true belief. Because neither the degree of one's own remorse nor the score of one's confessions can guarantee in any way the effectiveness of a confession ... True belief is therefore the only thing that is absolutely required. If it is there, all that is needed is a general confession containing at the most the revelation of only those sins that weigh most heavily on one's conscience ... But someone who does not have the right attitude of the heart and beliefs, and the true desire for salvation, should not go to confession. The fact that the church requires confession may not cause him to confess. On the contrary, he should stay away from it until he comes to a better understanding and the right attitude of the heart.

(1902, p. 82)

During Luther's absence from Wittenberg, this view led to consequences that the reformer had not intended at all. It was Carlstadt in particular who had a hand in abolishing confession over a period of several years. As soon as Luther returned from the Wartburg toward the end of 1523, he immediately complained about this development and prepared for the reintroduction of confession in a new form. Luther thought that it was scandalous, the number of 'undeserving' who came to the Last Supper. He introduced (besides the confession that was voluntary) a kind of examination of belief to which everyone who wanted to be admitted to the Last Supper had to submit.

After this more catechetic and dogmatic examination, those who were deemed to need it were subjected, if necessary, to an ethical religious examination. In it, the priest had to pay attention to the individual's conduct of life. If the priest found someone who did not satisfy the requirements in that respect or had obviously committed sins, then he had to carefully investigate whether the person had stopped committing sins, or was at least deeply distraught about his sins and trying to break away from them.

This bisection of voluntary confession into an inner religious act and the policing of morality by the church had very important consequences.

The inner self-direction through consciousness was combined very effectively with external control by strangers. After all, in a Protestant community, to be excluded from the Last Supper because of an immoral lifestyle was not only a religious matter, but also had very serious consequences for the individual's secular status. Generally speaking, however, the examination of one's beliefs did not affect everybody equally. Fischer points out that questions were only asked once, and not at all of 'persons of high intellect or morality'. Thus there were two types of social control: on the one hand, social control, the effectiveness of which was based on conscience, was for those who were experts in religious matters, and for well respected members of the community; on the other hand, there was externally based social control for the rest of the people.

The 'discovery' of intentions in the twelfth century led to a new understanding of action, but on its own was not sufficient to give rise to a biographical perspective. While it is true that since the twelfth century individual acts were no longer viewed in isolation from the corresponding motivations, there was no effort to analyze an individual's single acts in the context of all his acts. Still lacking was the idea that there is a connection between the biographical context of the act and the whole of the actor's life as an individualized system of processes that are interconnected by intentions. Church confession (at least that of the lay-person and, possibly to a lesser degree, that of the monk or other members of the church) even had a side-effect: acts that were confessed and regretted could be expunged from memory for, after all, they are expunged from God's 'memory' since He has forgiven them. In this sense, confession tied the sinner to single acts, emphasized the connection between motive and behavior, and developed the idea of individual responsibility; but all this was done without reference to the individual's life as a whole. If confession took away the burden of one's past, it could not produce any impulse to systematize all of life's components. This would require a new step in the understanding of behavior. It is possible that the systematization of ascetic life among European monks brought with it something like a responsibility that results from life as a whole. This form of relating life and behavior only transpired during the Reformation, and was then also required of laymen; Max Weber[14] saw this most clearly.

He said that when salvation no longer depends on single deeds, and when the saving Grace cannot repeatedly be lost through sin and regained through confession, but instead becomes a 'sentence' in the form of predestination hanging over one's life as a whole, then the individual certainty of being saved (*certitudo salutis*) can no longer be gained from singular acts but must become a reflection of one's biography as a whole.

Therefore the question cannot be: through which deeds, words, and thoughts have I insulted God? Instead, it must now be: is my life as a whole that of someone who belongs to the chosen? As a consequence, life must be subjected to systematic control. The merely sporadic confession would be a much too unsystematic instrument of regulation. Max Weber points out that in relation to the older form of Catholicism, Calvinism is characterized primarily by the enormous strengthening of systematic control of behavior in all aspects of life (Weber, 1920/1956). Weber mentions different instruments that work toward this regulation. First, there is the new valuation of time. Time is sacred and must be used; no moment can be wasted. Now, even short moments of sin represent a waste of time, hint at one's possible damnation, and can no longer simply be expunged through confession and remorse. Weber mentions in this connection the ascetic principle of self-control that characterized the Puritan and 'made him the father of modern self-discipline' (Weber, 1920/1956, p. 117). However, the most important principle that characterizes the Puritan according to Weber, is the systematic control of affects, a methodical life in inner-worldly asceticism.

Weber is well aware that the principle of self-control was nothing new and developed step-by-step as an other-worldly principle in the lives of monks, and even in the lives of the laymen who emulated them. What is important is the redirection of the monks' morality into a requirement that became the duty of every layperson; it changed from an other-worldly directive for life that existed for a religious subgroup into a general rule. The Calvinist's certainty of having been chosen is not the result of 'periodic release from affective guilt' through confession, but rather the result of an alternative – to be chosen or condemned. It is astonishing that this description of the Puritan's life ideal which, when it is secularized clearly becomes the program for a process of civilization, has rarely been viewed from this point of view. The question that has generally been discussed is only whether a connection really exists between the ideals of self-control and the creation of the spirit of capitalism. Norbert Elias's theory of civilization continues to use descriptive categories that are parallel to those of Weber,[15] but he does not take into account the religious roots of the process of civilization. Weber had already pointed to the type of the courtly, civilized gentleman in the Anglo-Saxon civilization who 'values reserved self-control' (Weber, 1920/1956, p. 117) and is dependent on Puritan ideals. However, Weber scarcely touches upon the institutionalized form in which this new systematic self-control was being practiced.

Our thesis is that we have here new forms of confession that are of central importance. Weber, in his treatment of Benjamin Franklin, has pointed to his diary, but he concentrates only on its aspect of the budgeting of time. From our point of view, it seems important to interpret the institution of the diary as a type of confession that makes self-reassurance through biography possible. The connections between the Puritan diary, autobiography, and the development of the bourgeois novel in England are obvious, especially if one considers the works of Daniel Defoe or Samuel Richardson. The successor to the church confession within Puritanism was, in a sense, 'self-confession', but we should not forget that church confession does not simply disappear in Puritanism. Lewin Schücking (1929), for example, mentions that the Puritan marriage was also a confessional unit; religion would contribute to the couple's salvation and bind them together totally by giving them the opportunity of baring their souls completely. To this end, they form yet another communion within the communion of the family. The forms in which this confessional unit are manifest may seem peculiar to us because they not only include common religious interests, reading and meditation, but also a kind of service by the name of humiliation, where prayers are offered that are not only self-confessions, but also a listing, with a plea for forgiveness, of sins, weaknesses, and failures committed by the spouse, who on this occasion learns of them for the first time. To this are added prayers of gratitude for observed virtues in the spouse. In this way, the idea of confession changes in a peculiar manner (Schücking, 1929, p. 55).

Originally, the insistence on self-exploration and self-control resulted from concern as to the certainty of being saved. This led to the general rationalization of the conduct of life. According to Schücking, the connection is obvious between this incessant 'labor of introspection' and the increasingly refined psychological sensibility that characterized English novels at the time. The new time perspective is also clearly visible: on the one hand, there is the utilization of every moment and, on the other, the development of paying attention to one's biography as a whole, living according to a long-range perspective, and keeping a diary. As a result, the individual develops a heightened feeling for and consciousness of his own unique self.[16]

There are other reasons why the originally religious biographical perspective (encompassing life as a whole) appears in the bourgeois English novel of the eighteenth century, where the valuation of everyday and *inglorious* phenomena seems to be part of the theme, and these become acceptable as material for serious works. This change has to do with the Puritan ethos, leading to emphasis on the dignity of everyday life. It is not the one-time heroic act that can ensure salvation, rather the fact of belonging to the chosen ones is manifest in methodically fulfilling one's duty through the course of time. Puritan holiness, unlike Catholic holiness that can be interpreted as a case of withstanding extraordinary temptation, is seen as 'proving yourself with respect to minute detail', as heroism in everyday life. In other words, Puritan holiness is biography.

## Confession and 'general confession' during the French Counter-Reformation

What can be said of the development of confession in Catholic contexts during the same period? Sociology has hitherto largely neglected the modernizing influence of the Counter-Reformation. We will only provide some hints here. One of the most striking changes concerns the new understanding of confession, leading to practical effects similar to those in Calvinist countries, although the theological foundations were totally different. The starting point for Catholic thinking was the problem of the person who had confessed and repented his sins, and then relapsed into sinning. The new Catholic psychology, especially of the French Counter-Reformation was that the sinner could not have relapsed if he or she had really repented. Thus, presumably, the first confession was not really valid, and the confessed sins were not forgiven. The only psychological guar-antee of sincere repentance was for a person not to relapse. Ideally, only a permanently ethical, irreproachable life could be accepted as a psychologi-cally sufficient and authentic repentance in the single confession. Thus, just as in Calvinism, only the biography as a whole warranted salvation. The corresponding institutional device was the introduction of the so-called 'general confession'.

Norbert Elias argues that the methodical conduct of life and the strict self-control of modern, educated and polite people are the result of the power of monopoly of the court. From our viewpoint, the methodical con-duct of life can be seen (at least in the Puritanical sphere) as having a mass base resulting from new religious pressures and procedures encouraging the search and control of one's conscience. Sociological literature pays little attention to these phenomena and even less attention to the changes in mentality that accompany them in Catholic lands immediately after the Counter-Reformation.

The devotion of the layman leads to a reformulation of numerous exer-cises of piety. Accordingly, the pious layman should not castigate himself through fasting, but instead demonstrate the subjugation of his self by eating, without complaining, whatever is put on the table. In this way one can humiliate oneself without affecting anyone else. Only this kind of devotion is suited for life outside a monastery (see de Sales, 1969, p. 197).

However, in bourgeois life, this devotion should not be learned without guidance. The individual should not be left to his own devices but should put himself under the control of a spiritual guide, the 'directeur de l'âme'. This soul guide can, but need not, be the same person as the father confessor. In either case, the soul must be totally *bared* to him. Increase in self-control, the intensification of the method or conduct in life as required by devotion, occur step by step. Devotion is a slow process; it is like a career. It would be careless, even dangerous, to submit too quickly to the external control of the father confessor (see de Sales, 1969, p. 41).

It is tempting to misinterpret the role of the 'directeur de l'âme,' allowing him to appear as if he has to shoulder all the responsibilities. The opposite is the case. The father confessor is at the same time witness and judge, to whom the person confessing has to answer. Even the most fleeting thought and the most far-fetched act become intersubjective knowledge through confession; they become something real that cannot be doubted, an element of one's own biography that cannot be denied in its reality. In the devotion of the Counter-Reformation, the 'general confession' was the most important means for the objectification of one's own life course. This was a very old term, but it had meant something different in the Middle Ages. The medieval meaning is spelled out in a passage from Alain de Lille's *Summa de arte praedicatoria*. From it we learn that the concept of 'confession' has two meanings: it can mean either 'general confession,' or 'special confession'. The general confession may be made during the daily morning or evening mass; it refers to hidden and forgivable sins. The special confession is the auricular confession *per se*, during which the mortal sins that one knows of are confessed (see de Lille, p. 172f.). The general confession of the Middle Ages was not, therefore, a confession of concrete sins, but rather the admission in general that one is a sinner. One admits to having sinned, but not to what.

Differing from this, the concept 'general confession' during the Counter-Reformation meant a special confession concerning concrete sins. It did not mean the periodic listing of sins, but a one-time or rarely occurring confession that encompasses the entire past life; it was a biography of sins, so to speak. Here the candidate for devotion takes inventory of his entire past and recalls what was forgotten, in order to make it known to himself and to the father confessor. In a sense, the individual's biography becomes fixed. Even the forgiven, long confessed sins and one's inclination toward them, as well as the temptations that were overcome through the grace of God, are to remain elements of self-consciousness (see de Sales, 1969, p. 43). This continuous conscience-control was to continue after the general confession, if possible before the simple weekly confessions, and every evening during the private examinations of one's conscience (see de Sales, 1969, p. 95).

## Confession as modern biography generator?

The first question we must ask is whether confessions that contain the total biography have become dysfunctional in our society, since this is characterized as having differentiated systems such as law, economy, politics, religion, and private spheres isolated from spheres of public action. It appears, in fact, that the individual develops multiple biographies that correspond to the differentiation into subsystems. These biographies must be synchronized from time to time when incompatibilities and disturbances develop. The acts and motivations that enter these partial biographies depend on the criteria relevant to the sphere of life to which the individual is to be coordinated. The biography that is reconstructed from someone's medical history is based on factors that differ from those in the dossier representing the individual's *curriculum vitae* before an employer or the Secret Service. This difference in the selection of biographical data does not mean that one ought to restrict the given information to only that one sphere. The *curriculum vitae* required by an employer does not exclude private information. To the contrary! But in this context, it is not the whole of the private life that is of interest. To some degree, limits are established, that is, protection of privacy, secrecy of confession, secrecy of medical records, to prevent the partial biographies being recombined into the whole biography.

It is characteristic for present times that the information that one supplies about oneself when filling out a questionnaire, being interviewed by the authorities, during office hours or in one's *curriculum vitae* is not a revelation, which is then forgotten. Today there are numerous methods to store and preserve these disclosures, and to recombine them according to criteria that we cannot control. There are methods to analyze our disclosures for hidden patterns in order to derive from them insights into health, reliability, or mental capacity: medical tests such as the AIDS tests or blood examinations are just some ways of focusing on our body or physiology.[17] The storage of confessions that has become possible in unprecedented volume through modern electronic means represents a change, the consequences of which cannot yet be evaluated. All confessions have always been linked to an aspect of control, manipulation, and surveillance; but today this has reached an entirely new quality. The fear of such total surveillance already manifests itself in many publications. It can be found as the vision of 'Big Brother' and similar nightmares depicted in science fiction novels and films, as well as in serious scientific publications.[18] The utopia or nightmare of the 'transparent person'[19] would then represent the conclusion of a development that began with the individual's voluntary self-revelation before the priest, who is sworn to secrecy but represents an omniscient God; the individual must become transparent to the father confessor as he already is to his God. However, the father confessor took no notes and did not maintain electronic files; his knowledge could not be stored or centralized. In addition, he only had knowledge of a small group of people. The power of the father confessor depended primarily on the faith of the confessing person. In the modern fiction of the 'transparent person', the secularized father confessors would no longer be representatives of God; they would simply be omniscient.

There can be no doubt that the information gained through confessions and disclosures is relevant, even in the context of modern differentiated institutions. But we may ask – and here we come full circle – whether confessions or other forms of self-thematization[20] are still meaningful to the individual. If we analyze the literature of various periods, we have the impression that the individual used to be a unique whole that was, by and large, responsible for himself, but that today he has been divided into small pieces. It appears as if the past in its totality is no longer significant for the identity of the individual, but that the individual can, depending on the situation, evoke or forget certain elements. It may be possible that social change occurs so fast, and the systems to which one has to relate are so complicated, that the individual's personality and character as structures with firm content can no longer adapt over a long period of time. This was probably the thought behind Riesman's thesis of the 'other directedness' of modern individuals (Riesman, 1950). The multiplicity of groups we belong to makes it impossible to develop a permanent and unified self. If the individual belongs to only one group, 'this group not only knows his acts, it also confronts him with them and attributes them to him. Because the group permanently attributes them to him and reflects them, he learns to view himself as an actor with an identity, and attributes his acts to himself. It is this mirror of a tight group that creates a person's identity' (Tenbruck, 1963, p. 34). Today we are far from such a situation. We can no longer create a whole out of the stream of time that runs through our biography because there is no social entity confronting us that confirms its objectivity. Therefore,

> the identity of the individual is reduced ... When the adventures of the time we went to school, our adventures in college, on vacation, immediately are in the past and behind us because there is nothing tying these social realms together, and we can – so to speak – begin again; when marriage and entertainment fall into entirely separate groups; when we can at any time change our social group by changing our job

or moving, then our acts turn into 'what is happening' or 'what has already happened.' They have become something that can be looked at from the outside.

(Tenbruck, 1963, p. 34)

On the other hand, the large number of printings of conventional novels that include consistent biographies show not only that this kind of novel has not yet come to an end, but also that the idea behind it, namely the wholeness of a life course, has not yet totally lost its social currency. The fact that we no longer receive from the mirror of a social group an image of our past that cannot be reinterpreted arbitrarily and solipsistically does not mean that self-thematizations have become inconsequential. However, their character has changed considerably. The self that is thematized in this fashion is privatized to a previously unknown degree. This means that if it is at all binding, it is so only for its bearer, and possibly only to a degree. Within this private frame of self-thematization there are numerous techniques for the formation of identity. The private character is not negated by the fact that the agencies one might use for self-thematization are highly organized, often commercialized, and most often professional. Berger and Luckmann speak of 'identity markets'. There the classical religious techniques coexist with numerous new techniques that range from individual therapies to self-experimental groups.

The main function of these strategies of finding oneself seems to be not so much securing social control, but rather creating meaning: not so much assuming responsibility for sins, but producing happiness through overcoming trauma. The firm commitment to one's past is less important than the selective use of the past for 'explaining' current crises. Sometimes the goal is the symbolic orgiastic re-creation of trauma; in many cases, this may lead to 'overcoming' one's past. When the aim is primarily the synchronization of disparate experiences and contents of consciousness, this overcoming of one's past may, in certain cases, be achieved by forgetting certain episodes instead of confessing them. Rarely does the intention create a desirable biography that is harmonious; what is more likely to be at stake is the permanent redefinition of one's biography through new confessions. Here the criterion for selecting relevant pieces of the past is the present with its varying needs for meaning and catharsis. Sometimes, however, the harmony of the biography may be achieved without a reflexive technique, through direct self-experience in trance or ecstasy, through acting out, or through immunization against the past. If confession was once the technique for fixing the self to its contents, the new forms of confession serve more to energize the self in the face of externally produced adaptation. It has been said of totalitarian regimes that they constantly rewrite their history. This is also true for the modern individual and the contents of their confessions. New forms of self-thematization appear in public, like the increasing number of 'autobiographies' of public persons or Internet diaries[21] (like web logs with regular entries of the owner, some open to the public, some restricted to a certain group of persons chosen by the author, like friends, members of the family, colleges, communities of interest, etc.). This quite recent phenomenon corresponds to our thesis about the increasing need of man to find ways for self-thematization in modernity (in the sense of the widely systematic manner mentioned above). If we broaden our analysis of biography generators to other fields of empirical research, different types of qualitative interviewing methods, especially biographical in-depth interviews can also be consulted. These empirical methods are chosen when the exploration aims to obtain information about certain persons (often in the sense of 'typical cases') that demand more open techniques of investigation such as explorations about identity. To choose a more concrete example: types of transnational identity that cannot be clearly identified by closed questions which elude narrow 'either-or definitions' of national affiliation. Changing the definition of one's self is experienced as part of the autonomy of the individual who can interpret his life (more correctly, his private life) subjectively. To the degree that our self loses its objective and firm obligations, it becomes for us the narcissistic source for more, always new and interesting, novels.

As has been shown, changing structures of society correspond with different ways of self-thematization, ultimately based on the will to monitor and subject oneself to internal or external guidance. Both modern and Christian variants of self-thematization not only produce socio-specific forms of biographical identity, but also forms of social control. Social control is thus not antithetic to, but firmly based upon, self-thematization as a most prominent technology of the will.

## Notes

1. The discussion should be seen both in the context of the work of neurophysiologist Benjamin Libet (Libet, 1985) and in the philosophical facet of the topic represented by Popper and Eccles (1977).
2. Here we can refer to William I. Thomas: 'If situations are defined as real, they are real in their consequences.'
3. For a reflection of the actual debate about free will in media from a sociological point of view, see Maasen, 2006.
4. Concerning the debate in the Middle Ages, see von Moos, 2004.

5. Besides personal identity (on which we focus here) we can speak of another dimension as well: of *social identity*, which refers to the conscience of belonging to a certain group, structural characteristics, and the idea of membership. Both facets are closely linked with each other, are socially imparted (mainly in socialization) and constructed. Thus *every* form of identity is social, never completely independent of the surrounding social systems. Therefore the term 'social' identity is inadequate to describe all aspects. Consequently, Alois Hahn recommended another label in earlier works: 'participative identity' to underline the 'moment of collective identification and, if necessary, solidarity'. It also underlines the idea of construction: in using some characteristics for identification, the others take a back seat at first, but still remain. One can allude to it as 'plurality of functioning selves'. To illustrate that: for instance, the affiliation to a certain ethnic (or religious) minority group can be irrelevant for the self/self-thematization of a person in general, but can acquire an all-dominant feature in another situation (where the identification with a profession or nation etc. becomes less important). 'Participative identity' is primarily ther-referential, corresponds to social status, functions, or social groups to which the person dedicates (or in contrast does not dedicate), whereas 'biographical identity' is *stricto sensu* self-referential and can be defined by its emphasis on personal history in so far as it is not shared by others, focusing on one's own experiences, attributes, emotions, etc. Certainly, as the latter depends on the acceptance of others and the first is not naturally given but has to be acquired, identity is both other-referential *and* self-referential (see Hahn, 1999, p. 74).

   Erving Goffman points out that the central problem is to balance and coordinate personal and social identity. Both are ascribed identities that he contrasts to what Erik H. Eriksons called 'ego or felt identity': 'Social and personal identity are part, first of all, of other persons' concerns and definitions regarding the individual whose identity is in question ... On the other hand, ego identity is first of all a subjective, reflexive matter that necessarily must be felt by the individual whose identity is at issue' (Goffman, 1963, p. 129).

6. That this identification is not absolutely reliable and sometimes, depending on illusions or ambiguities, as well as the idea of uniqueness, can be distressed by doubles, twins, etc. is discussed in Hahn and Schorch, 2007.

7. For more detailed information, see Hahn, 2000a, p.99.

8. The 'Lebenslauf', life course as *total* of life's events, experiences, etc. has to be distinguished from 'biography' as *selection* of the 'Lebenslauf' and its thematization and realization. See ibid., p. 101.

9. For a good summary of confession in ancient Christianity and the Early Middle Ages, see Poschmann, 1930 and, in addition, Rahner, 1973.

10. Thomas Aquinas points out that the 'forma' of the confession, which consists of forgiving the confessed sins, relies on having been instituted by Christ (Thomas Aquinas, 1938, p. 7).

11. Because one could hardly imagine a human being who never commits a deadly sin, and deadly sins can only be expunged through confession (besides baptism), confession is necessary for salvation (see Thomas Aquinas, 1938, p. 5).

12. The relevant texts can be found in Sess, 14 c. 4. A textbook summary can be found in Noldin, 1914, pp. 296–315.

13. A summary can be found in Dietterle, 1903–1907; see also Tentler, 1977.

14. Here we only treat the development of the process of civilization on the basis of religiously founded psychogenetic processes, using Weber's essays on Protestantism, because he was the first (together with Ernst Troeltsch) to describe this phenomenon precisely. Of course, this should not replace a thorough study of the sources from this point of view, nor should it prevent the discussion of the relevant counter-arguments against Weber's theses. For a summary of criticism on Weber's essay and his replay, see Weber, 1968.

15. Such as principles of internalization of self- and affect-control, the revaluation of time and the all-embracing rationalization of the biography (see Elias, 1939).

16. Not all life memories or *memoirs* are diaries or autobiographies in the sense intended here. Hartman Leitner (1982, p. 113) showed that the systematic self-thematization of one's life is a recent phenomenon. Older 'autobiographies' are more likely to emphasize the congruency between the author and his social position.

17. Another example is the widely discussed 'naturalization test' in Baden-Württemberg which examines the attitudes and sincerity of foreign persons who intend to acquire German citizenship. That inner attitudes and intentions not only of others, but also of ourselves, are never completely accessible, was already mentioned above. For more details about tests and their exclusionary function, see Hahn and Schorch, 2007.

18. This is probably most apparent in the works of Michel Foucault. By using many frameworks ranging from the hospital to the prison and from the insane asylum to psychoanalysis, he traces the connection between the individual being controlled by information about himself that he provides during voluntary or forced discourse and a new form of social power (see Foucault, 1963, 1972, and 1975). All these cases are about securing information by putting individuals in institutions. But this changes when the control of impulses occurs as a result of voluntary discourse. It is this aspect that is central to Foucault's work on the history of sexuality (1976). There, what is said about sexuality could be said about practically all spheres of behavior (see ibid., p. 35).

19. Not fictional at all, but very real methods of observation take over more and more space in our lives, such as observation cameras in public areas, recording of personal data on credit and customer cards or biometric identification.

20. For theoretical implications of this concept, see: Luhmann, 1973, pp. 21–46. Here, however, the 'reflexive' element is transferred from the individual to the social system. Remaining unanswered is the question whether the insights that are gained during the analysis of the self-thematization of social systems can enhance the individual's understanding of socially institutionalized self-thematization.

21. In this context, we could place attempts to establish modern forms for confession provided by the Internet, such as 'inner computer contemplation' and the opportunity to confess sins online.

## References

Abelardus, P., *Scito te ipsum (Ethica)*, ed. V. Cousin, Petri Abelardi Opera t. II (Paris, 1836).

Aquinas, Th. v., *Summa Theologiae* (1265/66-1273), ed. J. Bernhart, vol. 3 (Stuttgart: Kroner, 1938).

Biel, G., *Collectorium circa quatuor libros sententiarum* (Lugduni, 1514), xx6b.

Browe, P., 'Die Pflichtbeichte im Mittelalter', *Zeitschrift für katholische Theologie*, LVII (1933): 335–83.

Dietterle, J., 'Die Summae confessorum sive de casibus conscientiae von ihren Anfängen an bis zu Silvester Prierias (unter besonderer Berücksichtigung ihrer Bestimmungen über den Ablaß), Untersuchung II: Die Summae confessorum des 14. und 15. Jahrhunderts bis zum Supplementum des Nicolaus ab Ausmo', *Zeitschrift für Kirchengeschichte*, XXVII (1906): 166–71.

Elias, N., *Über den Prozeß der Zivilisation. Soziogenetische und psychogenetische Untersuchungen*, vol. 2 (Basel: Haus zum Falken, 1939).

Fischer, E., *Die Geschichte der evangelischen Beichte*, vol. 1 (Leipzig: Dieterich, 1902).

Rotterdam, E. v., *De libero arbitrio diatribe sive collatio*, Ia1, Ausgewählte Schriften, ed. W. Welzig, vol. 4 (Darmstadt: Wissenschaftliche Buchgesellschaft, 1969).

Foucault, M., *Naissance de la clinique. Une archéologie du regard médical* (Paris: Presses Universitaires de France, 1963).

Foucault, M., *Histoire de la Folie à l'age classique* (Paris: Gallimard, 1972).

Foucault, M., *Surveiller et Punir. Naissance de la Prison* (Paris: Gallimard, 1975).

Foucault, M., *Histoire de la sexualité*, vol. 1: *La Volonté de savoir* (Paris: Gallimard, 1976).

Goff, J. le, *Pour un autre Moyen Age. Temps, travail et culture en Occident: 18 essais* (Paris: Gallimard, 1977).

Goff, J. le, *La Naissance du Purgatoire* (Paris: Gallimard, 1981).

Goffman, E., *Stigma: Notes on the Management of Spoiled Identity* (Englewood Cliffs, NJ: Prentice Hall, 1963).

Hahn, A., 'Identität und Selbstthematisierung', in A. Hahn and V. Kapp (eds), *Selbstthematisierung und Selbstzeugnis: Bekenntnis und Geständnis* (Frankfurt/M.: Suhrkamp, 1987), pp. 9–24.

Hahn, A., 'Soziologische Aspekte von Geheimnissen und ihren Äquivalenten', in A. and J. Assmann (eds), *Schleier und Schwelle. Archäologie der literarischen Kommunikation V*, vol. 1 (München: Wilhelm Fink, 1997), pp. 23–40.

Hahn, A., 'Eigenes durch Fremdes. Warum wir anderen unsere Identität verdanken', in J. Huber and M. Heller (eds), *Konstruktionen Sichtbarkeiten* (Wien and New York: Edition Voldemeer Springer, 1999), pp. 61–87.

Hahn, A., 'Partizipative Identität', in A. Hahn, *Konstruktionen des Selbst, der Welt und der Geschichte. Aufsätze zur Kultursoziologie* (Frankfurt/M.: Suhrkamp, 1997/2000a), pp. 13–79.

Hahn, A., 'Biographie und Lebenslauf', in A. Hahn, *Konstruktionen des Selbst, der Welt und der Geschichte. Aufsätze zur Kultursoziologie* (Frankfurt/M.: Suhrkamp, 1988/2000b), pp. 97–115.

Hahn, A., 'Luhmanns Beobachtung der Religion', *Kölner Zeitschrift für Soziologie und Sozialpsychologie*, LIII (2001): 580–9.

Hahn, A., 'Geheim', in G. Engel, K. Reichert and H. Wunder (eds), *Zeitsprünge. Forschungen zur Frühen Neuzeit*, vol. 6 (Frankfurt/M.: Klostermann, 2002), pp. 21–42.

Hahn, A., *Erinnerung und Prognose. Zur Vergegenwärtigung von Vergangenem und Zukunft* (Opladen: Leske and Budrich, 2003).

Hahn, A. and M. Schorch, 'Tests und andere Identifikationsverfahren als Exklusionsfaktoren', in Volkswagenstiftung (ed.), *Grenzen, Differenzen, Übergänge: Spannungsfelder inter- und transkultureller Kommunikation* (Bielefeld: transcript, 2007, forthcoming).

Leitner, H., *Lebenslauf und Identität. Die kulturelle Konstruktion von Zeit in der Biographie* (Frankfurt/M. and New York: Campus, 1982).

Libet, B., 'Unconscious Cerebral Initiative and the Role of Conscious Will in Voluntary Action.', *The Behavioral and Brain Sciences*, VIII (1985): 529–39.

Lille, A. de, *Summa de arte praedicatoria*, PL 210.

Luhmann, N., 'Selbstthematisierung des Gesellschaftssystems', *Zeitschrift für Soziologie*, II (1973): 21–46.

Maasen, S., 'Hirnforscher als Neurosoziologen? Eine Debatte zum Freien Willen im Feuilleton', in J. Reichertz and N. Zaboura (eds), *Akteur Gehirn – oder das vermeintliche Ende des handelnden Subjekts. Eine Kontroverse* (Wiesbaden: VS Verlag, 2006), pp. 287–303.

Mead, G. H., *Mind, Self and Society: From the Standpoint of a Social Behaviorist* (Chicago: University of Chicago, 1934).

Moos, P. von, 'Das Geheimnis der Prädestination im Mittelalter', *Internationale Zeitschrift für Philosophie*, II (2004): 158–92.

Noldin, H., *Summa Theologiae Moralis III, De Sacramentis* (Innsbruck: Rauch, 1914), pp. 296–315.

Popper, K. and J. C. Eccles, *The Self and its Brain: An Argument for Interactionism* (New York and Berlin: Springer, 1977).

Poschmann, B., *Die Abendländische Kirchenbuße im frühen Mittelalter*, Breslauer Studien zur historischen Theologie, vol. 16, ed. F. X. Seppelt et al. (Breslau: Müller & Seiffert, 1930).

Rahner, K., *Schriften zur Theologie, vol. XI: Frühe Bußgeschichte in Einzeluntersuchungen* (Zürich: Benziger, 1973).

Riesman, D., *The Lonely Crowd* (New Haven: Yale University, 1950).

Sales, F. de, *Oevres* (Paris: Bibl. Pléiade, 1969).

Schücking, L., *Die Familie im Puritanismus. Studien über Familie. Literatur in England im 16., 17., und 18. Jahrhundert* (Leipzig and Berlin: Teubner, 1929).

Simmel, G., 'The Sociology of Secrecy and of Secret Societies', *American Journal of Sociology*, XI (1906): 441–98.

Singer, W., 'Verschaltungen legen uns fest. Wir sollten aufhören, von Freiheit zu sprechen', in Ch. Geyer (ed.), *Hirnforschung und Willensfreiheit. Zur Deutung der neuesten Experimente* (Frankfurt/M.: Suhrkamp, 2004), pp. 30–65.

Singer, W. and G. Roth, 'Wir können nicht anders', *Frankfurter Allgemeine Zeitung*, CCLVI (4 November 2003): 33.

Tenbruck, F. H., 'Kultur im Zeitalter der Sozialwissenschaften', *Saeculum*, XIV (1963): 25–50.

Tentler, T. N., *Sin and Confession on the Eve of Reformation* (Princeton: Princeton University, 1977).

Weber, M., *Gesammelte Aufsätze zur Religionssoziologie*, vol. 1 (Tübingen: J. C. B. Mohr, 1920/1956).

Weber, M., *Die protestantische Ethik II. Kritiken und Antikritiken*, in J. Winckelmann (ed.), (Hamburg: Siebenstern, 1968).

# Part II
# Self and (Socio-)Scientific Knowledge

## Introduction

Addressing self and (social) science, this part reflects on the role of knowledge in practices and problematizations that ascribe central value to the will. Psychopharmacology and sociology are being reframed, not least to account for willing selves.

As Max Weber has shown, scientific progress must not be equated with a general increase in knowledge on conditions of life. Using a tramway, the individual usually neither knows nor needs to know its workings, as Weber pointed out. The same is true for use of supersonic transport, computers, or scanning tunneling microscopes. While the savage surpasses the modern individual in terms of knowing his tools, he is far from commanding a rational account of the world. What is lacking is precisely what Weber defined as the core of rationalization by way of science and scientifically oriented techniques: it is the conviction that 'principally there are no mysterious incalculable forces that come into play, but rather that one can, in principle, master all things by calculation' (Weber, 1919/1973). This 'disenchantment of the world' (Weber), being the flip side of the Enlightenment, positions human beings as masters over nature. To Max Horkheimer and Theodor W. Adorno, who have pointed to the irrationality in rationalization concomitant to the Enlightenment, man's domination of nature is threefold: it encompasses not only outer nature, but also inner nature and the domination of others (Horkheimer and Adorno, 1944/1994). In all these regards, the development of the respective knowledge led to the development of corresponding technologies rendering nature, ourselves, and others calculable.

Coming from a different theoretical tradition, Michel Foucault has directed attention to the relevance of contacts between self-technologies and technologies of domination in terms of power/knowledge. He dispenses with the notion of a subject of knowledge and the notion of ideology at the same time:

> We should admit rather that power produces knowledge that power and knowledge directly imply one another; that there is no power relation without the correlative constitution of a field of knowledge, nor any knowledge that does not presuppose and constitute at the same time power relations. These power-knowledge relations are to be analyzed, therefore, not on the basis of a subject of knowledge who is or is not free in relation to the power system, but, on the contrary, the subject who knows, the object to be known and the modalities of knowledge may be regarded as so many effects of these fundamental implications of power-knowledge and their historical transformations.
>
> (Foucault, 1975/1977, pp. 27–8)

To Foucault knowledge is to be seen as a process altering the subject: precisely by entering a field of knowledge, selves constitute themselves according to a fixed and determined status (Foucault, 1978/1996, p. 52).

When we refer to such a concept of power/knowledge with all its implications, human ontology proves to be a constantly contested object. Whereas technologies of internal monitoring render individuals governable in terms of calculability, responsibility, and self-regulation (see Part II, 'Self – Past and Present' in this volume), the psy sciences have provided (material) technologies for externally monitoring individuals, their actions, and motives. Inquiring into recent developments within the neurosciences, Nikolas Rose shows in Chapter 3 how their account of human mental life imposes extensive requirements and obligations on the individual by way of developing new technologies, such as brain imaging, molecular neuroscience, psychopharmacology, and behavioral genomics. Dispensing with individual biography and experiences as key factors of a willing self, neurosciences identify bodily factors for the explanation of human behavior. Consequently, changing one's behavior becomes a question of operations performed on the body, the brain, that is. Far from merely opening up an opportunity to apply novel self-technologies, the scientific knowledge provided turns into an obligation to make adequate use of this knowledge in terms of governing ourselves.

Whereas for Rose the inquiry into biomedicine reveals it to be a pivotal site for the 'fabrication of the contemporary self', Armin Nassehi in Chapter 4 analyzes the role of willing selves in different functional subsystems such as medicine, mass media, art, religion, law, politics, education, and science. Although different technologies of the will may be at work in the respective domains, one general function of employing will is identifiable: each subsystem makes use of the idea of free will for the sake of its own stability. Social systems address willing selves by way of communication that ascribes relevance to human beings and their agency. Strikingly, sociology, too, with little questioning, relies on the notion of an autonomous subject, even though the social conditions of autonomy have been an issue of concern since its inception. This leads Nassehi to conclude that especially sociological research treats as a solution what should be its genuine problem: how the actor becomes an actor in the first place and why modern society counts on the free will of individuals.

In this perspective, practices of attribution can be read as technologies of the will that are of vital significance to society. The sociology of regimes of freedom, discipline, and control has to scrutinize the accounts of individuality at stake. Relying on a concept of power/knowledge, the task is to reveal the co-constitution of fields of knowledge and power relations pivotal to the analyses of neoliberal societies, the subjects they call for, and the technologies they imply. It is a complex of self-technologies, technologies of domination, and material devices that affords willing selves and allows for the self-regulation of individuals by freedom in a society of control.

Throughout these regimes of knowledge – here pharmacological or sociological – the actors and their agency are both subject and object of expectations as to how they should regulate themselves and each other so as to be assigned the status of a morally and politically responsible actor. In a way, by continuously forming and reforming the norms of expectable behavior, the actors become objectified as voluntarily acting by subjecting themselves to self-technologies, technologies of domination, and material devices. The latter spin a tight web of practices and rationalities that do not converge into a fixed set of notions or norms. Rather, they give rise to a regime of calculability (Covalevski et al., 1998). It is complicit in individuating actors as well as rendering them 'autonomous' and 'responsible' according to 'rational(ized)' schemes of possible action. A regime of calculability subjects actors to a general rule of perception, thought, and action – which is one of controlling oneself and others. As methods of governance in a wide variety of societal domains are turning to voluntaristic and market-based forms of control, reflexive and locally flexible control and information structures have developed that markedly rely on self-governable actors whose modes of governance comply with the (flexible, yet rational) regime of calculability.

# References

Covalevski, M. A., M. W. Dirsmith, J. B. Heian and S. Samuel, 'The Calculated and the Avowed: Techniques of Discipline and Struggles over Identity in Six Big Accounting Firms', *Administrative Science Quarterly*, 43, 2 (1998): 292–327.

Horkheimer, M. and T. W. Adorno, *Dialektik der Aufklärung. Philosophische Fragmente* (Frankfurt/M.: Fischer, 1944/1994).

Foucault, M., *Discipline and Punish: The Birth of the Prison* (New York: Pantheon, 1975/1977).

Foucault, M., *Der Mensch ist ein Erfahrungstier. Gespräch mit Ducio Trombadori* (Frankfurt/M.: Suhrkamp, 1978/1996).

Weber, M., 'Wissenschaft als Beruf', in M. Weber, *Gesamtausgabe*, vol. 17 (Tübingen: J. C. B. Mohr, 1919/1973), pp. 71–111.

# 3
# Governing the Will in a Neurochemical Age[1]

*Nikolas Rose*

The idea of the will has led a strange existence. The *Oxford English Dictionary* traces its first use in the sense of 'desire, wish, longing; liking, inclination, disposition (*to do* something)' to Beowulf in the seventh century and suggests that in modern usage this sense is merged with another – that of 'the action of willing or choosing to do something; the movement or attitude of the mind which is directed with conscious intention to (and, normally, issues immediately in) some action, physical or mental; volition' which it first identifies in the Old English of the tenth century. In the nineteenth century, matters of the will were central to philosophy, to the emerging discipline of psychology, and to those concerned with the practical arts for the management of conduct, from pedagogy to passion. Yet today, while the notion of 'free will' remains a topic of debate in philosophy and in jurisprudence, the notion of the will itself has largely disappeared from the language of the disciplines of the subjective.

Yet, despite this disappearance of the will from the truth discourses of the psy sciences, the will lives on within the hybrid field of thought and action concerning the government of the self (Valverde, 1998). In this field, 'the will' remains an essential referent for all endeavors that seek to control impulses in the name of civility, for all modes of judgment that seek to hold the individual responsible for his or her actions, and for all those ways of relating to the self that we have come to understand as freedom. A person's 'will' is something that can be invoked in accounting for all kinds of conduct, both desirable and undesirable, and in doing so, it locates its origin within the individual actor. Will, in the context of the government or self-government of conduct, is something that can be acted upon: shaped, strengthened, molded or broken, tamed, disciplined, controlled. Despite the fact that its psychological coordinates remain unclear or

81

nonexistent, the will thus constitutes a kind of irreal space where responsibility, volition, culpability and much more can be located.

In the nineteenth century, the will was, at one and the same time, something given, in an inherited constitution, something shaped by experience, something modified by the very habits that it itself gave rise to, and something that was subject to action upon, by self or by others. Has anything changed in the new field of truth about human beings at the start of the twenty-first century? And if it exists, does the will have a seat in body or brain?

Over the last 50 years of the twentieth century, psychiatry gradually mapped out what it considered to be the neuronal and neurochemical bases of human mental life. We have witnessed the birth of biological psychiatry as a new 'style of thought'. This new way of thinking not only establishes what counts as an explanation – it establishes what there is to explain. The key here is a particular organ – the brain. Of course the idea of the brain as an organ that is the seat of thought, consciousness, conscience, intention and memory is an old one. But the brain now appears in a radically new way: visualized at a molecular level and anatomized in terms of molecular activities. For the psy sciences of the twentieth century, the organic brain was often merely the physical substrate within which the psychology of the human being was located – a deep psychological space with its own systems, rules, processes, within which the personhood of each individual was located, and which was the key to understanding thought and action, emotion and volition. But this deep psychological space that opened in the twentieth century has flattened out. Psychiatry no longer distinguishes between organic and functional disorders: the Cartesian dualism of body and soul seems to have been overcome. Mind is simply what brain does. And mental pathology is simply the behavioral consequence of an identifiable, and potentially correctable, error or anomaly in some aspect of the brain. This is a shift in human ontology – in the kinds of persons we take ourselves to be. It constitutes a new way of seeing, judging and acting upon human normality and abnormality. This shift enables us to be governed in new ways. And it enables us to govern ourselves differently.

It is possible to identify four key interlinking dimensions of this mutation: brain imaging; molecular neuroscience; psychopharmacology; and behavioral genomics. Each has implications for the government of conduct, linked in different ways to questions about the voluntary nature of action, notably criminal behavior, addiction and other forms of undesirable conduct that are seen to be linked to an inability of the individual to exercise appropriate controls over his or her thoughts, words and action. Let me say a few words about each.

## Brain imaging

Electron microscopes were invented in the 1930s: first transmission and then scanning. Brain tissue could now be imaged at resolutions 1000 times greater than those possible with visual microscopy – but only post-mortem. The properties of the skull ensured that the brain remained largely resistant to visualization with X-rays and other early techniques (Kevles, 1997). But a rapid series of inventions in the last part of the twentieth century overcame this barrier: Computerized Tomography (CT) in the 1960s, Single Photon Emission Computer Tomography (SPECT), Positron Emission Tomography (PET), Magnetic Resonance Imaging (MRI) from the 1980s and now Functional Magnetic Resonance Imaging (fMRI). It seems that we can now visualize the interior of the living human brain and observe its activity in real time as it thinks, perceives, emotes and desires. We seem to be able to see 'mind' in the activities of the living brain. Hence, many suggest that we are able to use these images of brain activity in different regions to make objective distinctions between normal and pathological functioning. Such claims have led to huge investments in scanning apparatus in medical and research facilities, and to the increasingly routine use of such scans in medical procedures and research. The significance of this apparent visualization of mind is not simply rhetorical – or even clinical. It is epistemological. The visualized living brain appears to be just one more organ of the body to be opened up to the eye of the doctor. When mind seems visible within the brain, the space between person and organs flattens out – mind is what brain does.

From images of the craving brains of drug addicts to those of the hallucinating brains of people with schizophrenia, it now appears that one can actually see emotions, desires, thoughts, intentions, and feelings in the living brain. One example will have to suffice: a study entitled 'Does Rejection Hurt? An fMRI Study of Social Exclusion' carried out by Naomi I. Eisenberger, Matthew D. Lieberman and Kipling D. Williams, published in *Science* in October 2003. This was a neuroimaging study that tried to examine the neural correlates of social exclusion and sought to test the hypothesis that the brain bases of social pain are similar to those of physical pain. Participants were scanned while playing a 'virtual ball-tossing game' in which they were ultimately excluded. The authors report that, paralleling results from physical pain studies, the anterior cingulate cortex (ACC) was more active during exclusion than during inclusion and that the activation of this region correlated positively with self-reported distress. Further, the right ventral prefrontal cortex (RVPFC) was active during exclusion and correlated negatively with self-reported distress. They argue that ACC changes mediated the RVPFC-distress correlation, suggesting that RVPFC regulates the distress of social exclusion by disrupting ACC activity. And they conclude: 'This study suggests that social pain is analogous in its neurocognitive function to physical pain, alerting us when we have sustained injury to our social connections, allowing restorative measures to be taken. Understanding the underlying commonalities between physical and social pain unearths new perspectives on issues such as why physical and social pain are affected similarly by both social support and neurochemical interventions ... and why it "hurts" to lose someone we love' (Eisenberger, Lieberman and Williams, 2003, p. 292). Emotional hurt consequent upon social encounters is not just metaphorical hurt, rejection hurts because it 'really' hurts in the body – in the brain.

## Molecular neuroscience

But what is 'the brain'. The last half of the twentieth century saw a shift from an electrical to a chemical understanding of brain processes. The initial focus was on synapses – the junction points between neurons – and the neurotransmitters that operate to transmit signals from one neuron to another across these junctions. The first neurotransmitters identified were the monoamines (dopamine, norepinephrine, epinephrine, acetylcholine, and serotonin), later some amino acids were also found to be neurotransmitters (notably gamma aminobutyric acid or GABA) and by the start of the twenty-first century, the number had grown into the hundreds. In addition, as molecular neuroscience developed finer and finer techniques for mapping the brain at this molecular level, the image of the functioning brain became more complex. Many new entities and processes were now involved in neurotransmission: receptor sites, membrane potentials, ion channels, synaptic vesicles and their migration, docking and discharge, receptor regulation, receptor blockade, receptor binding among them. These entities and processes were first hypothetical, then demonstrated in the lab, then became part of the routine explanation of the results of experiments (the sequence nicely characterized in Hacking, 1983). Finally, they became realities that seemed to be visible in their own right, independent of the visualization techniques that rendered them: stable enough to become more or less uncontroversial and conventionalized images in neuroscience textbooks (e.g., Stahl, 1996).

Such entities are now the common currency of the explanatory language of neuroscience, and by the end of the twentieth century it was in these terms that disorders previously attributed to failures of will are explained. Consider, for example, this press release on November 1999 from the

National Institute on Alcohol Abuse and Alcoholism (NIAAA) describing research published in *Nature Neuroscience* (Lewohl et al., 1999):

> Neurobiologists from the Waggoner Center for Alcohol and Addiction Research and Section on Neurobiology, and the Department of Pharmacology and Toxicology, College of Pharmacy, University of Texas (UT) at Austin discovered actions of alcohol while studying a subtype of potassium channels, a diverse family of ion channels that perform many central nervous system functions. Identification of the alcohol-sensitive channel has significant implications because of its key role in brain function. 'Molecular analysis of this cell membrane channel ultimately will increase our knowledge of how alcohol affects the brain and, thereby, the way a person functions,' said Enoch Gordis, M.D., Director, National Institute on Alcohol Abuse and Alcoholism (NIAAA), a component of the National Institutes of Health. 'The cellular effects of alcohols on the central nervous system have significant implications for understanding alcohol addiction,' said R. Adron Harris, Ph.D., study coauthor and Director of the Waggoner Center at the University of Texas. 'We have begun genetic manipulation of the new membrane channel to determine how it influences alcohol consumption and dependence.'[2]

Note here the form of explanation – alcohol affects the brain via its action on specific ion channels, and this affects the way a person functions – it is now at the level of the brain and its molecular processes that matters previously ascribed to the will, such as alcohol consumption and dependence, are to be understood, and hence it is at this level that they are to be treated. And while Professor Adron Harris refers to the genetic bases of variations at this level, it is not through genetics that treatment is to be attempted – it is through drugs.

## Psychopharmacology

Contemporary psychiatric drugs are widely believed to have their effects because they have been configured at the molecular level to have a calculable and specific action on anomalies within the neurotransmitter system – targeting receptors, ion channels, and other specific sites. At least in the early phases of the development and use of the first of this new generation of drugs, the Selective Serotonin Reuptake Inhibitors for the treatment of depression, it was suggested that this 'targeting' would enable normalization of neurotransmitter action, and thus rectification of the symptoms of the anomaly, with a minimum of the unwanted 'side-effects' that were associated with an earlier generation of 'dirty' drugs with a wide and unspecific spectrum of action. Eli Lilly's Prozac (fluoxetine hydrochloride) was the prime exemplar, but it was soon joined by a whole range of other drugs for the treatment of mild to moderate depression, and later for a host of other disorders: Generalized Anxiety Disorder, Pre-Menstrual Dysphoric Disorder and so on. A lucrative market has been carved out for these drugs, and they have become among the best selling pharmaceutical worldwide, with the United States leading the way (Rose, 2003, 2004).

This optimism has been punctured by time. Most specifically, it has been undermined by the recent acceptance by drug regulatory agencies in many countries that the SSRI class of drugs do indeed carry risks of adverse effects, sometimes severe, including risks of self-harm and suicide, and that these risks have been exacerbated by the very widespread pre-scribing of these drugs as a first line of treatment for many conditions, with the prescription of such drugs for children being a particular area of concern. Warnings have been issued, and general practitioners have been urged to utilize other treatments first, and not to prescribe certain of the drugs to children at all. The status of these drugs has been further undermined by evidence showing that the claims for efficacy of the drugs were considerably overstated, and that these claims seemed plausible only because the pharmaceutical companies have concealed the results of unfavorable trials, and published only those studies that appear to demonstrate the superiority of their own drug (Whittington et al., 2004; Whittington et al., 2005).

Nonetheless, more generally, the modulation of mood and desire by psychopharmaceuticals has become routine (Rose, 2003, 2004). This is linked to new modes of introspection, new ways of thinking about and understanding mood, will, and desire, new ways of charting variations in these aspects of our personhood, identifying anomalies and interven-ing upon them (Martin, forthcoming). They have led some to suggest that these new psychopharmaceuticals allow the calculated engineering of mental states – that they actually or potentially allow human beings to move 'beyond therapy' – that is to say the normalization of an abnormal or pathological state – to the manipulation of normality itself – to what is often termed 'enhancement'. These hopes and fears over enhancement technologies have led to a wave of publications in the field that is now known as neuroethics (Marcus and Charles A. Dana Foundation, 2002; Caplan, 2003; Marcus, 2003; Moreno, 2003; Wolpe, 2003; Kennedy, 2004; Sententia, 2004; Illes, 2005).

However, there is little evidence to suggest either that these drugs are either able to manipulate mental states at will, or are used in attempts to do so (Rose and Singh, 2006). Older drugs – such as cannabis and alcohol – or certain illegal drugs synthesized specifically for their rapid mood-changing effects – such as MDMA (methylenedioxymethamphetamine or Ecstasy) – are much more effective, immediate, and more widely used for such purposes. And psychopharmaceuticals are not promoted, marketed or prescribed on such a promise. In fact the rhetoric of the drugs is quite the opposite. They do not promise a 'new you' or the capacity to 'design your moods,' but are framed within a rhetoric of authenticity that is little different from the ethic that underpins other psychological therapies based on talk (Rose, 1989). Of course, there remain many locales and practices in which psychopharmaceuticals are used, under actual or virtual compulsion, by authorities in order to manage conduct that they consider undesirable, in prisons, psychiatric hospitals, and for those with mental health problems being treated 'in the community'; enhancement, here, is hardly at issue. Similarly, such drugs are often used by authority figures – parents, schoolteachers – in the management of the conduct of children – the use of methylphenidate (Ritalin) or amphetamine compounds (as in Adderall) for attention deficit hyperactivity disorder being the obvious example. But outside these practices, in the routine treatment of mental troubles, what they offer to those who take them is the capacity, through drugs, to become yourself, to 'get your life back,' to 'feel like yourself again'. That is to say, these drugs promise to help the individual him or herself, in alliance with the doctor and the molecule, to discover the intervention that will precisely address a specific molecular anomaly at the root of something that personally troubles the individual concerned and disrupts their life, in order to restore the self to its life, and to make the person who consumes the drug an actor in his or her own life.

## Behavioral genomics

Across the twentieth century, and with increasing frequency in the post-eugenic genetics that took shape from the 1950s onward, many attempts were made to identify 'the gene for' depression, schizophrenia, alcoholism, aggression, homosexuality and other modes of conduct deemed pathological. Many bold claims were made by those who thought they had discovered such genes, all of which failed to stand the tests of replication or to demonstrate robustness in other populations. However, since the discovery that the human genome contains less that 30,000 coding sequences, not the 100,000 previously believed, the idea that there is one gene for each characteristic, disease or form of conduct is being abandoned (Venter et al., 2001). The search now takes a different form – it is a search for 'susceptibilities'. That is to say, researchers now try to identify variations in the human genome at the single nucleotide polymorphism (SNP) level – places in the genome where one of the four bases that make up the 'genetic code' – conventionally represented by C (cytosine), G (guanine), A (adenosine) and T (thiamine) – has been substituted by another – so that a sequence changes from CAGT to CAAT for example. It is now thought that such SNP-level variations, in combinations, may provide protections, modulate responses to the environment, or increase liabilities to develop particular characteristics in particular conditions.

It is in these terms that a cascade of papers have been published on susceptibility loci in ADHD, impulsivity, aggression, alcoholism, depression, schizophrenia, and much else. And it is in these terms that new forms of intervention are being imagined, such as the screening of children or offenders for the susceptibility loci in question and the use of preventive interventions, usually involving psychiatric drugs, to reduce the risk of the undesirable consequences. Of most interest to us in this present context, one key focus of this research has been upon a phenomenon quite close to the older idea of the will – that of 'impulse control' – the ability of an individual to exercise control – that is, restraint – over violent, aggressive or otherwise undesirable impulses – from outbursts of temper or other undesirable emotions, through gambling and 'addictions' to those involving a physical assault on another person.

Consider, for example, the following paper from 2000 entitled 'A regulatory polymorphism of the monoamine oxidase-A gene may be associated with variability in aggression, impulsivity, and central nervous system serotonergic responsivity':

> This study presents preliminary evidence of an association between polymorphic variation in the gene for monoamine oxidase-A (MAOA) and inter-individual variability in aggressiveness, impulsivity and central nervous system (CNS) serotonergic responsivity. An apparently functional 30-bp VNTR in the promoter region of the X-chromosomal MAOA gene (MAOA-uVNTR), as well as a dinucleotide repeat in intron 2 (MAOA-CAn), was genotyped in a community sample of 110 men. All participants had completed standard interview and questionnaire measures of impulsivity, hostility and lifetime aggression history; in a majority of subjects (n = 75), central serotonergic activity was also assessed by neuro-psychopharmacologic challenge (prolactin

response to fenfluramine hydrochloride). The four repeat variants of the MAOA-uVNTR polymorphism were grouped for analysis (alleles '1 + 4' vs. '2 + 3') based on prior evidence of enhanced transcriptional activity in MAOA promoter constructs with alleles 2 and 3 (repeats of intermediate length). Men in the 1/4 allele group scored significantly lower on a composite measure of dispositional aggressiveness and impulsivity (P < 0.015) and showed more pronounced CNS serotonergic responsivity (P < 0.02) than men in the 2/3 allele group ... We conclude that the MAOA-uVNTR regulatory polymorphism may contribute, in part, to individual differences in both CNS serotonergic responsivity and personality traits germane to impulse control and antagonistic behavior.

(Manuck et al., 2000)

Baroque as this form of reasoning may appear, it is in fact entirely characteristic of this style of thought. There is no claim about the discovery of 'the gene' for aggression, but of the identification of variations at the SNP level in a genetic sequence involved in the synthesis of an enzyme, monoamine oxidase A (or MAOA), involved in the metabolism – in this case, the breakdown – of a monoamine that acts as a neurotransmitter in the brain, serotonin (5-hydroxytryptamine or 5-HT) – that has been associated with variations in impulse control in adult subjects. In a 2003 review of research of this sort entitled 'Is There a Genetic Susceptibility to Engage in Criminal Acts?', Katherine I. Morley and Wayne D. Hall concluded:

Genetic research is beginning to identify genetic variants that may have some bearing on an individual's liability to develop antisocial behavioural characteristics ... This review of genetic research on antisocial behaviour has summarised growing evidence for a genetic contribution to antisocial behaviour but it has also indicated that it is highly unlikely that variants of single genes will be found that significantly increase the risk of engaging in violent behaviour. Instead it is much more likely that a large number of genetic variants will be identified that, in the presence of the necessary environmental factors, will increase the likelihood that some individuals will develop behavioural traits that will make them more likely to engage in criminal activities.

(Morley and Hall, 2003)

Not a 'gene for' then, but an interaction of multiple genetic variations that increase risk. What, then, are the implications for control practices of this new style of genetic thought?

## Genomics and crime

Writing in 1995, David Wasserman concluded that no mainstream researchers believe that there are single genes that cause violent or anti-social conduct; that all regard behavioral phenotypes like criminal behavior as arising from a complex interaction of many genes and environmental factors (1995, p. 15). He argued that none believe that genetic influence makes criminal behavior less mutable, and many suspect that the most effective ways of countering genetic influence will involve social and economic reforms. Finally, he claimed, few of these researchers advocate, or believe, their findings would support mandatory screening, involuntary medication, or harsher sentences. Wasserman was writing in a volume of papers deriving from a rather controversial conference on genetics and crime held in Maryland, and arising out of research funded by a grant from the Ethical, Legal and Social Implications (ELSI) branch of the National Institute for Human Genome Research (the papers are now published in Wasserman and Wachbroit, 2001). The conference was particularly controversial because, in a criminal justice system in which young black men are massively overrepresented in the courts, prisons, and probation, any suggestion that there may be a genetic basis for crime opens itself to the accusation of racism. Biological criminology had fallen out of fashion, and earlier claims, such as the XYY arguments that associated some male violence with possession of an extra Y chromosome, had proved to be based on flawed evidence and withdrawn (Saulitis, 1979). Yet, across the 1980s, some researchers held firmly to the belief that advances in molecular genetics, when linked to contemporary neuroscience, would have profound implications for the understanding of criminal conduct.

At the time of this conference, those working in this field had been particularly impressed by a study published in 1993 in *Science* by Han Brunner and his colleagues, who were examining one particular family in the Netherlands (Brunner et al., 1993a). In this paper, entitled, 'Abnormal-Behavior Associated with a Point Mutation in the Structural Gene for Monoamine Oxidase-A', Brunner et al. (1993b) described a 'large kindred' in the Netherlands where a number of young men exhibited a syndrome of borderline mental retardation linked to abnormal behavior, including violence and aggression They assigned the genetic defect for this condition to a particular region of the X chromosome, in the vicinity where it was 'known' that the genes for monoamine oxidase A and monoamine oxidase B were located. These enzymes are known to metabolize serotonin, dopamine and noradrenaline (known in the US as norepinephrine). Numerous earlier studies had reported findings of alterations in the metabolism of serotonin, and to a lesser extent dopamine and noradrenaline, in 'aggressive behavior in animals and humans,' for example, reduced concentrations of a breakdown product of 5-HT (5-HIAA) in cerebro-spinal fluid of those who have been designated as showing impulsive aggression.

Brunner and his team cultured skin fibroblasts from subjects who were affected and not affected, and showed that those from the affected males produced only negligible amounts of MAOA. They sequenced the relevant segment of the gene and identified four base substitutions in the affected gene – three of which were neutral polymorphisms (so designated on the basis of existing knowledge on coding sequences) but one of which changed a glutamine (CAG) codon to a termination (TAG) codon. This C to T mutation was found in each of the five clinically affected males, and in the two other non-affected heterozygotes that were also known to carry one copy of this gene. So it seemed that the behavioral syndrome – borderline mental retardation and a tendency towards aggressive outbursts, often in response to anger, fear, or frustration varying in severity and over time – was a result of the deficiency in MAOA, which was itself a result of a mutation involving a single nucleotide at a specified point on a particular chromosome. Brunner and his team concluded that 'Taken together, data obtained in this family suggest a relation between isolated complete deficiency of MAOA activity and abnormal aggressive behavior in affected males' (Brunner et al., 1993a, pp. 579–80).

Brunner and colleagues hedged their conclusions with qualifications: Would this relation be shown in other families? Are all the biochemical deficiencies related to MAOA deficiency relevant or only some? How frequent is MAOA deficiency in the population? Is aggressive behavior confined to complete MAOA deficiency? How can one explain the fact that MAO inhibition in adult humans has not been reported to result in an increased liability to impulsive aggressive behavior? Can studies of aggression in male rats help? Is there a mediating factor, for instance the effect of MAOA on REM sleep? But what does not seem in doubt is that there is something called aggressive impulsive behavior, it is induced, potentiated, or made more likely by brain neurotransmitters and all that might affect them, and that the system is controlled in some way or other by genes and the proteins they do or do not synthesize.

## The gene for aggression?

This paper by Brunner and his team has become something of a 'gold standard' for the explanatory form of the new behavioral genomics. It appears to find a direct sequence from a single nucleotide base change to pathological conduct, a chain that runs as follows: base substitution → altered enzyme → effect on neurotransmitter linked to conduct → pathological conduct. It has not only sparked off a philosophical and jurisprudential debate about its implications for 'free will' and responsibility, it has also been cited in mitigation by defense lawyers in murder cases in the United States. But before we rush to the conclusion that something has fundamentally altered in our understanding of these matters, we should pause. Even in the United States, I know of no case where a court has accepted scientific research of this type in a successful defense or plea for mitigation of sentence. Courts have consistently argued that, whatever the scientific status of these findings, they need to be shown a clear causal link between a specific biological or medical condition suffered by an accused at the time of the offence and the specific criminal act in question. And in no case have these requirements been satisfied. Further, Brunner himself has argued strongly that his research does not support any claim that there is a 'gene for aggression'. Writing in 1996, he insisted that his study gives no support for the notion of an 'aggression gene,' despite having been interpreted in this way by the popular press:

> the notion of an 'aggression gene' does not make sense, because it belies the fact that behavior should and does arise at the highest level of cortical organization, where individual genes are only distantly reflected in the anatomical structure, as well as in the various neurophysiological and biochemical functions of the brain ... although a multitude of genes must be involved in shaping brain function, none of these genes by itself encodes behaviour.
>
> (Brunner, 1996, p. 16)

Indeed, Brunner is not alone in making this type of argument. My favorite example is a 1995 paper by Heikki Vartiainen entitled 'Free will and 5-hydroxytryptamine' (Vartiainen, 1995); 5-hydroxytriptamine (5-HT) is the neurotransmitter serotonin, and his paper reports abnormalities in serotonin levels in the cerebrospinal fluid (CSF) of people who have taken their own lives and in offenders who have carried out violent crimes, some of whom show excessive sensitivity to alcohol. It links this to reports of a hereditary factor in some alcoholic and violent offenders; brain scans showing abnormalities in individuals with neuropsychometric deficit; arguments that impulsive aggressive individuals have decreases in certain aspects of the serotonin uptake mechanism in the medial prefrontal cortex. And it concludes that not all biological factors are hereditary, but nonetheless 'The relationship between a low serotonin turnover and impulsive, aggressive behavior seems to be obvious ... A display of uncontrolled and uncharacteristic anger following minimal provocation can be biologically explained – a decrease in brain 5-HT manifesting itself as aggressive behavior' (Vartiainen, 1995, p. 7). The end of free will? No! Vartiainen continues that 'Since all behavior is biologically based, attributing causation to a given type of conduct as biological and calling it therefore an illness, tells us nothing about the social, moral, or legal implications which that behavior ought to have' (1995, p. 8). And he concludes that sentencers should not be concerned with whether a biological condition weakened legal responsibility, but with the protection of society and the reduction of the likelihood of recurrence of violent acts.

Indeed, Vartiainen grasps the central function of the idea of free will in jurisprudence since at least the mid-nineteenth century. When the judiciary defend the non-genetic, non-psychiatric fictions of free will, autonomy of choice, and personal responsibility, this is not because legal thought considers this a scientific account of the determinants of human conduct. Legal thought finds it necessary to proceed as if it were, for different reasons – reasons to do with prevailing notions of moral and political order and the need to find those who commit acts culpable and liable to punishment, despite the fact that their acts might, in another form of thought, be capable of explanation in terms of factors ranging from the environmental to the biographical, and even the biological. Indeed the trend of legal thought seems to be in the other direction – especially in the United States. That is to say, it seems to be gradually moving in favor of imposing an inescapable moral responsibility on the offender, no matter what might account for his or her actions, and hence imposing the culpability that comes along with that. In this form of thought, no appeal to biology, biography, or society will be sufficient to weaken the moral responsibility for the act, let alone the requirement that the offender be liable to control and/or punishment.

The argument from biology is likely to have its most significant impact, not in diminishing the emphasis on free will necessary to a finding of guilt, but elsewhere – in the determination of the sentence. For if anti-social conduct is indelibly inscribed in the body of the offender, reform appears more difficult, and mitigation of punishment inappropriate: the possibility therefore is a call for long-term pacification of the biologically irredeemable individual in the name of public protection. And, indeed, in several states in the United States this is very evident. For example, in 1996 the Oregon State constitution was changed to shift the focus of criminal punishment from 'the principle of reformation' to 'the protection of society': a shift that was used by the judge in rejecting genetic evidence in the controversial case of Kip Kinkel, who had killed his parents and opened fire on children and teachers in the classroom of a school from which he had been expelled.[3] In the United States, and in many other countries, faced with what is perceived as a threatening epidemic of violent crime and by repeated panics about such monstrous individuals as serial killers, psychopaths, and sexual predators, the courts increasingly insist that moral culpability, especially in relation to violent or anti-social conduct, should not be mitigated by any factors – be they social, familial, medical, or biological.

We may, as Nietzsche predicted in 1878, have come to recognize that 'freedom of the will is an error'.[4] But we cannot, it seems, abandon the idea of responsibility. On the contrary, within the criminal justice systems of our contemporary cultures of individual accountability, we reconceptualize offenders as creatures inescapably required to bear full responsibility for the outcomes of their actions, and deem these actions to be moral choices whatever their material causes.

## But what of the will?

Nor, it seems, can we abandon the search for the neurogenetic basis of the will. In 2003, Evan Deneris and his colleagues at Case Western University, working with mice, reported the discovery of the Pet-1 gene – only active in serotonin neurones – which when knocked out produced elevated aggression and anxiety in adults compared to wild type controls (Hendricks et al., 2003). Their paper was entitled, modestly, 'Pet-1 ETS gene plays a critical role in 5-HT neuron development and is required for normal anxiety-like and aggressive behavior'. However, the Press Release from Case Western University, the location of the research in question was less modest. It was entitled 'Researchers Discover Anxiety And Aggression Gene In Mice; Opens New Door To Study Of Mood Disorders In Humans'.[5] The Press Release helpfully informs readers that 'Serotonin is a chemical that acts as a messenger or neurotransmitter allowing neurons to communicate with one another in the brain and spinal cord. It is important for ensuring an appropriate level of anxiety and aggression. Defective serotonin neurons have been linked to excessive anxiety, impulsive violence, and depression in humans ... Antidepressant drugs such as Prozac and Zoloft work by increasing serotonin activity and are highly effective at treating many of these disorders ... .' And Deneris himself comments: 'The behavior of Pet-1 knockout mice is strikingly reminiscent of some human psychiatric disorders that are characterized by heightened anxiety and violence.'

But if arguments of this kind are not going to destroy the fiction of free will in the courtroom, how will they impact on the criminal justice system, and on control mechanisms more generally. The answer, I think, is to be found in the centrality of ideas of risk and precaution in such strategies, at least in advanced liberal democracies. Consider a paper by Diane Fishbein, of the US Department of Justice, entitled 'Prospects for the Application of Genetic Findings to Crime and Violence Prevention' (Fishbein, 1996). Fishbein outlines a program of research to assess the relevance and significance of genetic findings for crime and violence prevention. She concludes that 'Studies suggest that a subgroup of our population suffers from genetic vulnerabilities that overwhelm most environments' (1996, p. 93). But, for her, this is a cause for optimism, for 'genetic traits are not immutable, they are alterable in a social environment ... Not only do these individuals stand to benefit greatly from the research, but the public may eventually give way to more tolerance of behavioral aberrations, understanding that behavior is not entirely volitional at all times in all individuals ... there is little evidence that present tactics are effective; thus we need to move forward into an era of early intervention and compassionate treatment that genetic research may advance' (1996, p. 93).

What is in prospect, then, is less dramatic, but perhaps more pervasive, than the loss of the idea of free will. It is the development of programs of screening to detect individuals carrying these markers. It is the introduction of pre-emptive intervention to treat the condition or ameliorate the risk posed by the affected individual. And the most likely route to be followed will involve the use of psycho-pharmaceuticals to reduce risk. Full-scale screening of the inhabitants in the inner cities might be too controversial to contemplate in most jurisdictions. But the example of Attention Deficit Hyperactivity Disorder, at least in the United States, suggests the possibility of genetic screening of disruptive schoolchildren, with pre-emptive treatment a condition of continuing schooling. Or one might imagine post-conviction screening of petty criminals, with genetic testing and compliance with treatment made a condition of probation or parole. Or one can envisage scenarios in which screening and therapy are offered to disruptive or delinquent employees as an alternative to termination of employment. Many psychiatric medications, for example antabuse for alcoholics and lithium for manic depression, were introduced in this way in the United States. Further, biological expertise might be called upon to screen for genetic markers and neurochemical abnormalities, in order to evaluate the levels of risk posed by offenders, or non-offenders with a mental illness diagnosis prior to discharge from prison or hospital. Release would be dependent on compliance with a drug regime. In the context of proposed changes of legislation in relation to violent offenders, sexual predators and other risky individuals in many jurisdictions, this scenario is hardly science fiction!

## Governing oneself: somatic individuality

The sense of ourselves as 'psychological' individuals developed across the twentieth century – we came to understand and act upon ourselves as beings inhabited by a deep internal space shaped by biography and experience, the source of our individuality and the locus of our discontents. Today, this is being supplemented or displaced by what I have termed 'somatic individuality' (Novas and Rose, 2000). By somatic individuality, I mean the tendency to define key aspects of one's individuality in bodily terms, that is to say, to think of oneself as 'embodied,' and to understand that body in the language of contemporary biomedicine. To be a 'somatic' individual, in this sense, is to code one's hopes and fears in terms of this biomedical body, and to try to reform, cure, or improve oneself by acting on that body. At one end of the spectrum this involves reshaping the visible body, through diet, exercise, and tattooing. At the other end, it involves understanding troubles and desires in terms of the interior 'organic' functioning of the body, and seeking to reshape that – usually by pharmacological interventions. While discontents might previously have been mapped onto a psychological space – the space of neurosis, repression, psychological trauma – they are now mapped upon the body itself, or one particular organ of the body – the brain.

Biology here is not destiny – the management of uncertainty is the norm, for molecular presymptomatic diagnosis gives no calculability to the 'when' or the 'how' of illness or death, the life decisions between the moment of diagnosis and this imagined end point. And biomedicine – however genetic and neurochemical – is a key site for the fabrication of the contemporary self: free yet responsible, enterprising, prudent, encouraging the conduct of life in a calculative manner by acts of choice with an eye to the future and to increasing their well-being individually, and in their communities of identity. In the world of susceptibilities, we are all asymptomatically ill: the ambit of disease, and the powers of the doctor extended to those who are neither phenomenologically nor experientially ill. While some critics imagine a future of anxiety, guilt, fear, stigmatization, and discrimination without hope of remedy, advocates claim that the diagnosis of genetic susceptibilities will help individuals take responsibility for management of their condition, to take control of their vulnerabilities through the use of individually tailored behavioral and pharmacological correction of underlying error or deficiency, enabling the susceptible individual to maintain themselves in their everyday life. This is not so much normalization as correction, not so much exclusion as responsibilization. It involves a new form of the prudential relation which we are obliged to take towards our future: genetic prudence, that is to say, new ways of crafting one's life and modulation of a lifestyle in the light of one's susceptibilities. In this new ethic, the potential sufferer is to become skilled, prudent, and active, an ally of the doctor, a proto-professional – and to take their own share of the responsibility for managing their biogenetic selves. Such a prudential norm introduces new distinctions between good and bad subjects of ethical choice and biological susceptibility. And it is linked, I suggest, to a more general shift in regimes of control, a move from implacable abnormalities to manageable susceptibilities, which is consistent with wider reshaping in practices for the government of persons.

## A society of control?

Gilles Deleuze, in a brief and speculative set of notes, suggested that contemporary societies are no longer disciplinary, in the sense identified by Foucault – they are societies of control (Deleuze, 1995). Where discipline sought to fabricate individuals whose capacities and forms of conduct were indelibly and permanently inscribed into the soul – in home, school, or factory – today control is continuous and integral to all activities and practices of existence. We are required to be flexible, to be in continuous training, lifelong learning, perpetual assessment, continual incitement to buy, to improve oneself, constant monitoring of health, and never-ending risk management. In these circuits, the active citizen must engage in a constant work of modulation, adjustment, improvement in response to the changing requirements of the practices of his or her mode of everyday life. In the dimension which I have been discussing in this chapter, we are each required, with the assistance of our authorities, to modulate our life in the light of a knowledge of our somatic individuality, an obligation that now extends to the government of the will itself.

## Notes

1 This paper was presented at 'On Willing and Doing: An international symposium in the framework of the interdisciplinary project on "Voluntary Action – On the Nature and Culture of Volition"', held at the Max Planck Institute for Psychological Research, Munich in February 2004. A more developed version of the arguments can be found in Nikolas Rose, *The Politics of*

*Life Itself: Biomedicine, Power, and Subjectivity in the Twenty-First Century*, Princeton University Press, 2006, and some passages in this chapter come directly from that book.

2  Accessed    February    2004    at:    http://www.niaaa.nih.gov/NewsEvents/ NewsReleases/discover.htm

3  Details of the trial and excerpts from transcripts of the evidence presented are at:   http://www.pbs.org/wgbh/pages/frontline/shows/kinkel/trial/   Accessed 29 August 2005. I discuss this case along with many others in Chapter 7 of *The Politics of Life Itself*, referenced in Note 1 above.

4  See Aphorism 39, 'The fable of intelligible freedom', in *Human, All too Human*.

5  Accessed 29 August 2005 at: http://www.sciencedaily.com/releases/2003/01/ 030123072840.html

# References

Brunner, H. G., 'MAOA Deficiency and Abnormal Behavior: Perspectives on an Association', in G. R. Bock and J.A. Goode (eds), *Genetics of Criminal and Antisocial Behavior* (Chichester and New York: John Wiley, 1996), pp. 155–67.

Brunner, H. G., et al., 'Abnormal Behavior Associated with a Point Mutation in the Structural Gene for Monoamine Oxidase-A', *Science*, 262, 5133 (1993a): 578–80.

Brunner, H. G., et al., 'Abnormal-Behavior Linked to a Point Mutation in the Structural Gene for Monoamine Oxidase-A', *American Journal of Human Genetics*, 53, 3 (1993b): 13.

Caplan, A. L., 'Is better best? A noted ethicist argues in favor of brain enhancement', *Scientific American*, 239, 3 (2003): 104–5.

Deleuze, G., 'Postscript on Control Societies', in G. Deleuze, *Negotiations* (New York: Columbia University Press, 1995), pp. 177–82.

Eisenberger, N. I., et al., 'Does rejection hurt? An fMRI study of social exclusion.', *Science*, 302, 5643 (2003): 290–2.

Fishbein, D. H., 'Prospects for the Application of Genetic Findings to Crime and Violence Prevention', *Politics and the Life Sciences*, 15, 1 (1996): 91–4.

Hacking, I., *Representing and Intervening: Introductory Topics in the Philosophy of Science* (Cambridge: Cambridge University Press, 1983).

Hendricks, T. J., et al., 'Pet-1 ETS gene plays a critical role in 5-HT neuron development and is required for normal anxiety-like and aggressive behavior', *Neuron*, 37, 2 (2003): 233–47.

Illes, J., *Neuroethics: Defining the Issues in Theory, Practice, and Policy* (Oxford and New York: Oxford University Press, 2005).

Kennedy, D., 'Neuroscience and Neuroethics', *Science*, 306, 5695 (2004): 373.

Kevles, B., *Naked to the Bone: Medical Imaging in the Twentieth Century* (New Brunswick, NJ: Rutgers University Press, 1997).

Lewohl, J. M., et al., 'G-protein-coupled inwardly rectifying potassium channels are targets of alcohol action', *Nature Neuroscience*, 2, 12 (1999): 1084–90.

Manuck, S. B., et al., 'A regulatory polymorphism of the monoamine oxidase-A gene may be associated with variability in aggression, impulsivity, and central nervous system serotonergic responsivity', *Psychiatry Research*, 95, 1 (2000): 9–23.

Marcus, S. (ed.), *Neuroethics: Mapping the Field: Conference Proceedings, May 13–14, 2002* (San Francisco and New York: Dana Press, 2002).

Marcus, S. J. (Ed.), *Neuroethics: Mapping the Field* (New York: Dana Press, 2003).

Martin, E., *Bipolar Explorations: Toward an Anthropology of Moods* (Forthcoming).

Moreno, J. D., 'Neuroethics: An Agenda for Neuroscience and Society', *Nature Reviews Neuroscience*, 4, 2 (2003): 149–53.

Morley, K. I. and W. D. Hall, *Is There a Genetic Susceptibility to Engage in Criminal Acts. Trends and Issues in Crime and Criminal Justice* (Canberra: Australian Institute of Criminology, 2003).

Novas, C. and N. Rose, 'Genetic Risk and the Birth of the Somatic Individual', *Economy and Society*, 29, 4 (2000): 485–513.

Rose, N., *Governing the Soul: The Shaping of the Private Self* (London: Routledge, 1989).

Rose, N., 'Neurochemical Selves', *Society*, 41, 1 (2003): 46–59.

Rose, N., 'Becoming Neurochemical Selves', in N. Stehr (ed.), *Biotechnology, Commerce And Civil Society* (New York: Transaction Press, 2004), pp. 89–128.

Rose, N. and I. Singh, 'Neuroforum', *BioSocieties: An Interdisciplinary Journal for the Social Study of the Life Sciences*, 1, 1 (2006): 97–102.

Saulitis, A., 'Chromosomes and Criminality: The Legal Implications of XYY Syndrome', *Journal of Legal Medicine*, 1, 3 (1979): 269–91.

Sententia, W., 'Neuroethical Considerations: Cognitive Liberty and Converging Technologies for Improving Human Cognition', *Annals of the New York Academy of Sciences*, 1013 (2004): 221–8.

Stahl, S. M., *Essential Psychopharmacology: Neuroscientific Basis and Practical Applications* (Cambridge: Cambridge University Press, 1996).

Valverde, M., *Diseases of the Will: Alcohol and the Dilemmas of Freedom* (Cambridge: Cambridge University Press, 1998).

Vartiainen, H., 'Free Will And 5-Hydroxytryptamine', *Journal Of Forensic Psychiatry*, 6, 1 (1995): 6–9.

Venter, J. C., et al., 'The Sequence of the Human Genome', *Science*, 291, 5507 (2001): 1304–51.

Wasserman, D., 'Science and Social Harm: Genetic Research into Crime and Violence', *Report from the Institute for Philosophy and Public Policy*, 15, 1 (1995): 14–19.

Wasserman, D. and R. Wachbroit (eds), *Genetics and Criminal Behavior* (Cambridge, Cambridge University Press, 2001).

Whittington, C. J., et al., 'Selective serotonin reuptake inhibitors in childhood depression: Systematic review of published versus unpublished data', *Lancet*, 363, 9418 (2004): 1341–5.

Whittington, C. J., et al., 'Are the SSRIs and atypical antidepressants safe and effective for children and adolescents?', *Current Opinion In Psychiatry*, 18, 1 (2005): 21–5.

Wolpe, P. R., 'Neuroethics of Enhancement', *Brain and Cognition*, 50 (2003): 387–95.

# 4
# The Person as an Effect of Communication*

*Armin Nassehi*

The subject of my chapter is the subject or its derivates, the actor, the author, the individual. But my aim is not to explain what a subject or its derivates are or can be or should be. The claim I assert is rather modest. I will not deal with real subjects – if any real subject has ever existed. I want only to meditate on how subjects can be addressed and why subjectivity as a form has emerged as an accountable point of attribution. I want only to make some theoretical remarks about how sociology deals with the problem of subjectivity. As a warning, I have to confess that I shall take readers on a necessary detour through the discipline's classical approach to theorizing subjectivity, with a focus on the problem of willing and doing and the deconstruction of those traditional concepts. The *basso continuo* of this chapter is a criticism of the peculiar tendency of sociology concerning the functional meaning of attribution practices to individuals. How selves can emerge as 'willing selves' is the fundamental question of this volume and the fundamental question of the bourgeois tradition. The basic question of this bourgeois world was: 'How can we will what we have to do?' The contemporary question in our neuroscientifical world is probably: 'How can we will what our brain is going to do with us?' Thus far – my cryptic suggestions about what I want to present here.

## The deconstruction of the subject as a phenotypical theoretical figure

It has become a phenotype to deconstruct the subject as an origin of the world or as an origin of action. So it is not really surprising that a sociological perspective turns around what is a matter of course in our everyday knowledge: that communication is an effect of persons and not the other way round. This technique of inversion of everyday plausibilities almost constitutes the state of a speaker in cultural studies or social science. And the arguments in this field are well known. There are different theoretical ways to come to that conclusion, but the decentralization of the strong autonomous subject has developed into a figure of thought which becomes regarded as a matter of fact. Both in literature and in cultural studies, both in postmodern and post-structuralistic philosophy and in social theory, the idea of deconstructing the actor and its derivates has become a shared conviction – and as for other phenotypical figures, their performance is related to a special kind of self-evidence which makes closer reasons unnecessary. Anyhow, the assertion of the death of the subject is a steadily asserted, seldom reflected pattern, which has comparatively little consequences for research practice in particular.

Although the world is populated by about seven billion subjects, intellectuals attempt to do their job of denying these billions of subjects their status as subjects. Doing that, these intellectuals demand to be the authors of the abolition of authorship, to be the subjects who decentralize subjectivities and to be the actors who ridicule the figure of an autonomous actor. What has become an expected attitude amongst intellectuals who have enough time at their disposal to stylize themselves as authorities in the literal meaning of this word seems to get refused by the practice of these authors who don't believe in authorship. Do they refuse practically what they have asserted theoretically? Do they make impossible what they do by doing it? Are they victims of the postmodern disease or of post-structural fever? Do they disclaim what makes them to be speaking while they are speaking?

Before I ask more questions like these, I should break here already at this early stage of my chapter. The point of my early break is to observe in detail what is going on here. My latter sentences can be decoded as the beginning of a well-known narrative. And as narratives tend to, they produce a special kind of internal dynamic. They force themselves to come to logical conclusions so one can say that the truth of narratives is their own practice. As a kind of self-developing structure, narratives supply communication with available patterns for redundant use. Given that, we should now have a closer inspection of how my argument has begun. Already the first sentences of my argumentation have brought up expectations about how the story will continue. If one begins with the paradox of subjects who reject subjectivity, or with authors who refuse authority, the narrative leads to arguments like those we know as the reproach of a performative contradiction of reasonable speaking: sentences that refuse their contents by their own practice are obviously wrong. Experts in that field of thinking can easily expect that the chapter will lead us to Jürgen Habermas's theory of communicative reason or to the demand for the avoidance of logical contradictions in rationalistic analytical philosophy.

And indeed: our postmodern and post-structural critics of subjectivity unintentionally give reasons about how to think about subjects, actors, and authors; not by *what* they usually assert but by *how* they do that. And from an empirical point of view, we have to notice that the social world is full of accountable targets we somehow treat as subjects. This is true, for example, for intellectuals who deny what they are practicing as accountable subjects, and it is true for me as a sociologist, writing a chapter that at this stage is very unclear whence it will eventually go, it is true for all those who can be made accountable, and – I am sorry – it is also true for all of you reading this text.

Let us once again inspect where the narrative has lead us. It would be possible and expectable to continue with a mitigation of the assertion that actors are not really actors, but effects of structures, of communication, or of social processes. But we could also recognize that the chapter has introduced lines of argument, slipping them in as narratives. This means that we can note a second performative contradiction. Even the critique of a contradictory assertion of the death of a reasonable speaker by reasonable arguments is pointing to a contradiction if we interpret this critique as part of a narrative. This story itself is a practice which produces speakers who are able to argue. What I want to emphasize is the following: If we do not account a story to a speaker but to the practice of the story itself, even a story about the autonomous subject is not more than a story. What we can learn from this is that the state of subjects, actors, and generally spoken accountable addresses depends on empirical accounts. Doubtless it is true that the critique of reason by reasonable claims for truth leads into a contradiction. But this is only true if we presuppose a familiar form of accountability which positions the speaker as the origin of communication, not as an effect of it. And there is no reason why not to prefer the latter. Probably, also, there is no reason for the former.

## The traditional semantics of subjectivity

I have to confess that this argumentation has led us into a dead-end street. But probably this is not a problem, but will turn out to be part of the solution I want to develop. What we have won so far is a hint at the practice of communication anyway. A solution cannot be found if we look for ontological reasons for what a subject or an actor or an author is, but by how such attribution targets can practically be set to work.

So I just want to begin once more. Supposably, it is surprising that I will now begin with the traditional concept of the subject, because I emphasize the practice of getting subjects into work, not theoretical concepts. But I want to propose to interpret the invention of the transcendental subject as a practice itself, a practice which reacts on fundamental societal changes.

The semantics of the subject and of subjectivity begins earlier than with transcendental philosophy. From Aristotelian metaphysics, the subject was something that was able to guarantee the existence of another thing or something that exists independently of something external. So it is not surprising that since the Cartesian *cogitans*, human consciousness has been advanced as the subject of the world. In Kant's philosophy, then, the subject was not the same as an empirical individual (Kant, 1996). Subjectivity was nothing individual, but universality *par excellence*; that means it was able both to abstract from empirical individuals and to connect them. In his philosophy of morality, Kant introduced the difference between reasonable beings and human beings, who are corrupted by addictions and needs. This differentiation makes visible the functional meaning of the theoretical technique on which transcendental philosophy is based. The transcendental, that is, non-empirical designation of a general subjectivity of reasonable beings, obviously corresponds to experience with unreasonable particular subjects, that is: with human beings. The mere intellectual idea of a self-acting, thinking subject for Kant is only a theoretical condition to reconcile the new social experience of freedom which has to be based in a structure which itself is not achievable for human freedom. I read this as a kind of a *theodicy of the free will*, as a theoretical attempt to reconcile the idea of universality and generality with the individual's willing. In other words: The transcendental figure of the reasonable subject is a reaction to the empirical unreasonableness of human beings. Their reason can appear only inside of themselves, but not in social reality.

It is easy to recognize that the invention of the reasonable subject is due to a social structure which changes from a traditional, stratified order to functional differentiation. The self-description of society has to underdetermine the individual because it has to change to a flexible self which is able to cope with changing and different social expectations. Individuals become included simultaneously in different social systems. Therefore bourgeois individuality arises as a focus which ultimately can emerge *beyond* those inclusions into several social relations and expectations. Parallel to the idea of a transcendental subjectivity, the ideas of inwardness and inner infiniteness emerge. As the bourgeois individual is compelled to develop a special kind of flexibility in its external relationships, in addition to this the changes and discontinuities of the outer circumstances have to be compensated for by an internal complexity which is able to both transcend the changing demands of the world and to compensate external diversity by internal unity. Therefore the individual comes to be the *subject* of its own subjectivity. The bourgeois individual then uses self-reference as the basic form of self-description.

In the bourgeois narrative of the autonomous individual, the combination of external under-determination and internal over-determination of the individual consciousness was compelled to *want* to be that sort of subject conceptualized in philosophy with its epochal semantics of subjectivity. The success of this semantics can be explained by its special potential to bring together the idea of reason with the unreasonable existence of the empirical individual. In terms of social practice, the idea of subjectivity can be understood only if we take into account the functional meaning of bourgeois self-description. Even though the semantics of subjectivity, of individual autonomy, and of free will were universalistic semantics beyond empirical outlooks, the ideas of freedom and equality were used distinctively to justify inequality. As the central motifs of bourgeois forms of self-description were education, self-control, sentiment, a special form of life-style, integrity of families, and later entrepreneurship and career-oriented goals for life, they symbolized a distinctive and stratifying strategy which used terms of humanity to describe only a small part of mankind. The most characteristic attribute of these new bourgeois forms of self-description is the harmonization of free will and self-control. What has been described as subjecting practices in prisons, in the military, in medicine, or in schools by Michel Foucault (1973, 1977) can already be observed as a frame of reference in the philosophy of subjectivity. The subject in that sense is autonomous only with regard to reason, and reason is the basis of a free will, but free will emerges as a problem when it can be limited by reason itself. So bourgeois autonomy has to be limited by reason, which itself is the precondition of autonomy.

It is this dialectics of particularity and universality that was the starting point of Hegel's criticism of Kant. For Kant, the other subject did not exist. The other subject was only a horizon of the empirical world. In philosophical terms, the other was saved in the categories in theoretical philosophy and in the principle of generalization in practical philosophy. Whereas Kant only used an abstract concept of reason to express the instance which stands for unity, Hegel had an empirical idea – in his early philosophy it was the acceptance of the other on behalf of social order (Hegel, 1979), in his mature thinking it is the idea of subordination under the objective spirit, historically symbolized in the epiphany of the state, *nota bene*: not as an abstract idea but as every empirical state (Hegel, 1967). In my opinion, Hegel's philosophy until today represents the central narrative of modernity which has to conciliate the freedom of the individual willing with its subordination under a generalized idea of order. The individual willing therefore is identical with its impossibility. In other words: The freedom of the free will is the insight of the necessity of a restriction of the individual will for the benefit of a generalized idea of companionship. I do not want to go into Hegel's attempt both to emphasize political companionship and the subordination of individual aspirations and to legitimate the economical sphere of the systems of needs, which he calls bourgeois society (*bürgerliche Gesellschaft*). But what has to be accentuated is the idea that in Hegel's thought, free will can only be regarded as really free if it is under the control of a generality which makes it part of something greater, finally part of an almost eschatological development of spirit.

I think it worthwhile to have a closer look at the philosophical tradition – not to enrich sociological concepts by philosophical ones, but to show that perhaps we often try to abolish a subject that never has existed. My short inspection has reflected both on Kant's idea of the subject as a theoretical precondition of thought and on Hegel's effort not to confront the free will with a subordinating social generality, but to conceptualize the free will *as* an act of subordination. Actually this short inspection shows that the semantics of subjectivity was able to harmonize the self-consciousness of a new bourgeois class with the restrictions of a new political order, a new order of labor, with new requirements of self-control, with education, with self-conscious morality and not least with authorities and authoritarian speaker's positions in society. I think that the idea of a free will and of subjectivity corresponds with a society in which both political power without physical violence and the subordination under the power of experts and professionals for *reasonable solutions* developed. The idea of a free will always refers to an asymmetrical structure in which there exist authorities for how to decide under free conditions. The emergence of professionals such as physicians, lawyers, priests, teachers, and other experts, provided with asymmetrical speaker's positions, does not mean a structure of command and obedience. The asymmetry these speakers can claim generates the conditions for insight and understanding, thus former recipients of commands come to be *subjects*; that is, subordinated under the demands of reason, which connect the freedom of the will with reasonable reasons for what has to be done.

Given all that, the traditional idea of subjectivity itself already implies its own deconstruction. The subject we today reject is a myth, nothing else. It is itself part of a practice which has helped to bring up the modern form of accountability we can deconstruct by intellectual means, but which we cannot abolish empirically. What we can do is to try to understand the functional meaning of this form of accountability and attribution to individuals. And if we speak about accountability and attribution practices we do not say anything about the ontological status of subjects, actors or authors, but we can reflect on how social practices generate attribution addresses (Nassehi, 2006, pp. 69–164).

## Sociology as an heir of the traditional philosophy of subjectivity

As I mentioned above, the deconstruction discourse does not really become noticed by sociology. This may have two reasons. On the one hand, sociology itself has begun as a deconstruction of the idea of an autonomous subject. On the other hand, sociology up to now has used the idea of autonomous action as one of its basic concepts. This seems to be a contradiction, but it isn't. The central frame of reference for sociology is the question how social order is possible even though individuals are free in their own will. Thus sociological perspectives start where the strong concept of subjectivity ends. They both have to emphasize the societal background of the empirical under-determination of the individual and describe this as a new form of social determination. A sociological perspective begins with the description of the conditions for this, what the philosophical tradition treats as its unconditional foundation. These conditions cannot be conceptualized as an *extra-social* instance, neither as reason nor as a historically developing spirit. Logically, early sociology seems to treat a similar problem as the philosophy of subjectivity does. It is a steadily asserted misunderstanding of the traditional concept that it conceptualizes subjects as autonomous and independent units. The author literally becomes *authorized* by external instances as a supra-individual reason, as an objective spirit or by a master knowledge which becomes enforced by professionals and other authorities, particularly reasonable authorities. Similar to this, early sociology conceptualizes society or social order as an authority which formulates the individual as a social actor. Consequently, for example, Emile Durkheim's idea of society is a compelling mechanism; Max Weber deconstructs subjectivity to a mere subjectively meant meaning which becomes enforced by its cultural significance; for George Herbert Mead, the self is a reaction on a significant or a generalized other; and for Talcott Parsons, the actor is only one part of a cybernetic system with internal dynamics and enforced especially by norms and the threat of sanctions. So a sociological perspective deconstructs the subjectivity of the subject, whether it wants to or not. I think sociology today is surprised about what remains of the subject in social space. What subjectivity then can be is dependent on the other side, and so the subject is always an effect of its counterpart.

Adorno has pointed out that for sociology the notion of *society* has succeeded the traditional concept of spirit (in a Hegelian sense) or generality (Adorno, 1972). What he wanted to emphasize was the context of sociological thinking: the experience of individual freedom and will on the one hand and of social necessity on the other. On this note, sociology appears to have inherited the bourgeois form of societal self-description which has to cope both with the experience of individual freedom and will, and with the experience of restricting forms of the individual will. The bourgeois character of this fundamental idea of sociological thought refers to two motifs: the first motif refers to the idea that everything that happens has to go through the heater of individual imagining consciousness. This I call the *virtualization* of the world, that is, the input of energy into the social world comes from the individual. This Kantian concept alludes to the fundamental bourgeois problem that the individual must will what it has to do. And, bringing us up to the present, since some fundamental papers by C. Wright Mills in the 1940s sociology can know that motifs and preferences are the results of attribution practices and, moreover, vocabularies and narratives about how motifs and preferences can be used (Mills, 1940). Sociology accurately knows that subjects have a *movens* which they are not themselves. What else but sociology could know that – probably with the exception of theologians? But sociology seldom draws the conclusion from this to go on to observe the effects of the functional meaning of attribution practices on motifs and preferences. On the contrary, sociology believes in motifs and preferences and treats them as quasi-natural sociological data.

The second motif also deals with the self-attribution of the bourgeois individual, but now more with the external origins of these motifs. Here the old Hegelian question arises of how atomistic individuals can become social individuals (Hegel, 1967). The well-known solution is the dialectical intermediation between the objective and the subjective sphere. From my point of view, social theory has never really become emancipated from this Hegelian heir in terms of conceptualizing society as a general sphere which is opposed to individuals whose aspirations have to become harmonized by subordination. Sociology, then, had two possibilities of using this theoretical figure: either as a criticism of societal expectations and limitations towards individuals or as a functional theory of normative integration. Above all, the latter has developed to the, so to speak, *normal science* part of sociology. The Hegelian national spirit, rendered in German as *Volksgeist*, as the empirical objective spirit, is the basic narrative for the sociological notion of society. So there can be observed a stable line from Hegel's authoritarian philosophy of the state to Durkheim's *conscience collective*, and Parsons's *societal community*, until Habermas's postulation of a *reasonable identity* of complex societies. In all these social theories the basic problem is how to deconstruct the individual will both as part and counterpart of a general sphere, called society.

Sociology seems to deal with an analogical problem like the traditional philosophy of the subject does. The subject with its own will, authorship, and freedom can only be thought of in the horizon of an instance which renders subjects not real subjects, that is, autonomous fundaments of the world. But sociology reverses the burden of proof. Whereas philosophy asks how freedom can be thought without a loss of the world outside, sociology's basic question is how to generate social order despite the fact that individuals are considered to be free. Both are peculiarly undecided, and both seem to avoid asking the really exciting question of how to observe the outer source of subjectivity.

In sociology, we can observe a peculiar *normalization* of this problem. To understand what makes scientific disciplines work, one should not observe innovative debates or scientific excellence. It is more interesting to look for those distinctions, notions, and arguments which work without a closer explanation. One has to observe what can be asserted in a scientific discipline with or without risk of contradiction. In other words: What is the basic narrative which forms the will of sociological scholars?

In sociology, this leads us to a very boring discourse about the different factors which contribute to action. Are there more individual or more structural contributions to action? Are subjective or objective parts of action predominant? Do we account social action to social systems or to individuals? The difference between a more structuralistic and a more individualistic approach is the origin of the two antagonistic tribes in social theory which have pacified their struggles by dividing the social landscape into different spheres of order: for example in micro and macro levels or in different spheres of aggregation. Nowadays the sociological warriors prefer a strategy of appeasement. Theories of social systems and of social action become harmonized in theoretical discourse. This policy of appeasement distributes the truth to both sides. The consequence are sentences like these: Okay, actions are both influenced by individual resources, capabilities, and energies on the one hand; and by social structures, expectations, and restrictions on the other. This is, of course, both a boring and a widespread solution, which amongst other things is the reason for the very sparse reception of theories of decentralization of the subject or of the actor in sociology.

Mostly we overlook the literal construction of this distinction between a subjective and objective sphere of social order, which emphasizes the individual as a source of contingency, energy, and creativity, whereas the opposite side becomes associated with limitation, restriction, and compulsion in a Durkheimian sense. Thus the sociological mainstream items deal with the individual as the more primordial ontological unit. The exciting result is that this is true not only for theories of action, but for the whole construction of normality in our discipline. The disastrous consensus between a more structuralistic and a more individualistic perspective cannot discern the functional meaning of an attribution to individual will because sociology itself uses this attribution technique for its own practice.

This may be an exaggerated picture of sociology. And, of course, it is if one only inspects theoretical confessions and introductory texts for teaching. But what is more important is the practice of research. And here sociologists apparently believe in actors and subjects. If they don't use aggregate data like socio-demographic statistics and so on, sociologists believe especially in communication which is effected by individuals. Both in qualitative and in quantitative research, individuals are the objects of questionnaires, interviews, and observations, they become observed as deciders and players, intentional actors, and subjects of their capabilities, and even this practice produces the state of accountable actors.

If sociology would observe its own practice, it could learn a lot about social practices. Questionnaires and interviews, the inquiry of opinions, and the idea of individual resources against social structures produce, literally invent, special forms of accountable and attributable persons. Such research practices themselves demonstrate that most of the practices in society are practices which do not come upon subjects, but which generate them. Free will, then, should not be observed as an individual resource, but as a part of social structure. We can only get over the simple distinction of individual creativity and social restriction if we look for the dynamics of the social itself. What we need is a praxeological theory which does not only observe individual practices within social structures, but observes how individuals emerge within social dynamics. To achieve an appropriate theory of individuality and its derivates – will, authenticity, authorship, subjectivity – obviously we need an operative theory of the social.

## A sociological alternative: inclusion theory

The decision for systems theory is not a decision for the social at the expense of the individual. For the theory of autopoietic social systems in the tradition of Niklas Luhmann, society consists only of communication processes which create structures by practical processes themselves (1986, 1990, 1995). The structural coupling of psychic systems and social systems does not describe a relationship of causality in one direction, but a relationship of reciprocity. There is no social structure outside of communicative practice, and this practice has to be described and attributed. And so we can sketch out that the kind of ascription and attribution to gods, human beings, organizations, groups, epochs or machines is contingent, that is, empirically manifold. A theory of this type works with only a few basic assertions. One of which is that all the narratives we use to make the world and what happens tellable are part of the social dynamics itself. So both the invention of the transcendental subject and the *normal* knowledge we have empirically about individuals and their free will is one of the problem-solving strategies in social systems.

As I have suggested above, the traditional idea of the subject is a reaction to a shift of the societal differentiation form. Whereas former societies did not have a use for what we today call a free will, the switch to the functional differentiation of society has to attribute more and more to individuals for reasons of its own order. I want to outline this in a bit more detail now – and I will eventually reach a point at which a theory like this also falls behind its own possibilities.

In the context of the theory of autopoietic social systems, the topos *individuality* appears in two regards. In the first instance, Luhmann speaks about the 'individuality of psychic systems' (Luhmann, 1995, p. 255) in the sense of an operative basis of operationally closed psychic systems. Their individuality results both from their operative closure and from the operative necessity to continue the connectivity of their own operations. This argument is reminiscent of Edmund Husserl's *Phenomenology of Inner Time Consciousness* (Husserl, 1991). Actually, Husserl has formed the paradigm of operational closure of event-based consciousness, a theoretical figure later called *autopoiesis*. In distinction to Husserl and the following phenomenological tradition in sociology, systems theory does not assume any kind of intersubjectivity, but the emergence of social systems as results, in this sense, of *individual*, that is, psychic closure. *Communication*, as the basic element of social systems, hence is not possible *although* psychic systems are not transparent to each other. Systems theory reverses this procedure: *Communication* as a system with its own internal logic and operational closure emerges *because* psychic systems are closed systems. In this respect, the individuality of psychic systems is the presupposition of sociology's fundamental question, how social order can be possible regarding the individuality of actors/human beings/psychic systems. This kind of *individuality* is unavoidable, and at best it is of interest for a basic theory of social systems.

More interesting for a theory of modern individuality is the question of how social systems create structural and semantic forms that effect special kinds of individualized ascription and inclusion. *Inclusion* stands for communicative strategies of considering human beings as relevant. *Inclusion* is the social mechanism that constitutes human beings as accountable actors, as *persons*. The concept of *inclusion* allows us to theorize the contact between social and psychic systems not as a kind of containment or membership, but as a mutual process of structural coupling. What is called *individuality* or *individualization* becomes the form of how persons cope with social expectations.

Social expectations depend on the primary differentiation scheme of society. In pre-modern, stratified order the inclusion of persons was actually organized after a pattern of containment and membership regarding to status, families, or regions. The whole stratified order followed a hierarchic code and so the inclusion of persons was easily paralleled to the entire society's basic structure. The individuality of persons was determined by the social position within this stratified order. The switch to functional differentiation now requires a stronger orientation toward self-reference. This applies both for the structure of societal differentiation and for the semantics of personal self-observation. The modern, functional differentiated society no longer follows one unifying code of self-observation. 'It multiplies its own reality' (Luhmann, 1995, p. 191). Modern society is at once the political function system, the economic function system, the scientific function system, the legal function system, the religious function system, and the educational function system with, in each case, their own respective environments within the societal system. Society gets decomposed in perspectives of function systems which cannot become organized or integrated as a whole. The main consequence of this theory is that functionally differentiated society has to be seen as a society without top or centre, that neither provides a coincidence of perspectives nor a perspective of coincidence. This kind of society renounces a strict regulation of system relations, replacing general intersystem relations with the relation between systems and environments or, in other words, replacing *strict coupling* with *loose coupling*. This makes even the recognition of unity within society difficult.

'The world can only be identified paradoxically on the basis of a particular code, that is, only as logically infinite information load. Furthermore, a distinction of distinctions, of coding and reference, remains possible. Society must be satisfied with this possibility and with the combinatorial latitude it provides. It can no longer refer to a final thought, to a reference-capable unity, to a metanarrative (J.-F. Lyotard) that prescribes form and measure. It is in precisely this sense that modernity's traditional semantics have failed' (Luhmann, 1998, p. 11). In other words: simulated central perspectives on the society have become replaced by *several central perspectives*. These perspectives are central and fundamental for themselves but they have to cope with other perspectives which cannot be reconciled with each other. This applies both to so called *cultural* perspectives provoking comparisons and the consciousness of contingency and to the invincible differences of the perspectives of function systems. Function systems are based on their own binary codes and their own functional perspectives, and this can be judged as their strong point and as their fatal limitation.

This limitation of function systems to their own codes and functional perspectives has a twofold effect on the structure of social expectations towards individuals. *On the one hand*, the function systems do not refer to individual wholeness. In their perspectives there exist only political, legal, economic, educational, or scientific expectations towards individuals, so that human beings are requested to take part in the different processes of these functional perspectives. *On the other hand*, a functional differentiated society depends on semantics of individuality that relieve the function systems from strict determination of individuality. From the perspectives of the function systems persons seem to be *dividuals*, divided by the uncoordinated access of the different functional perspectives on individuals (Nassehi, 2003, p. 100–25). But from the perspectives of the individuals, these divided perspectives have to be coordinated by themselves. This means that not only function systems become more and more related to a self-referential mode of observation, but also individuals. They have to handle this difference between *individuality* and *dividuality*, and that is what sociology usually calls *individualization* – this is, the spot where the problem of free will emerges.

In terms of systems theory, *individualization* is caused by the switch to functional differentiation. One can reduce this to the formula that premodern social order is based on strict membership and total inclusion, whereas modern society is built on multiple partial inclusions of the person. Corresponding to this, there emerge semantic figures of individual autonomy and social heteronomy, of freedom and alienation, of equality and inequality, of public and private spheres, of public welfare and self-ishness, and of uniqueness and standardization. These semantic forms are representations of the public battles which have been fought under the banner of *individualization* and *individuality*. Strictly speaking, this battlefield covers up a more exact sociological view on the *structural background* of the emergence of these central narratives of modern life. Perhaps it would be no exaggeration to say that this battlefield is only a secondary theatre, because these semantics of individualization as a central narrative of modern society are located in the *exclusion area* of modern society.

At least, this is the point of Luhmann's theory of individuality. He emphasizes that *individuality*, strictly speaking, is *exclusion* and that the self-observation of individuals is located outside of the function systems. This so-called *exclusion individuality* (Luhmann, 1989, p. 158) evacuates the conditions of self-reproduction from the function systems' perspectives and reduces the semantics of individualization to a kind of semantics besides the function systems. The sociological benefit of this concept doubtlessly is its sensibility towards the structural background of these semantics. Strictly speaking, this concept emphasizes the central paradox of modern individuality: a pattern that demands not to follow any pattern.

But that seems to be only a half-truth. Despite its theoretical benefits, I feel that the concept of exclusion individuality itself takes part in the *mythologizing* of the individual. Similar to Georg Simmel, this concept speaks of the individual as of something that can be achieved only in an *extra-social* sphere where the individuality could come to itself (Nassehi, 1999, p. 85). Even though Luhmann does not conceptualize individuality as an *extra-social* entity, he locates individuality in a social space that keeps distance from the function systems. *Exclusion individuality*, then, is a technique of distanciation, of self-segregation from the central units of society. And this seems to be strongly suggestive of the bourgeois individual. I do not criticize this, but I want to emphasize that the consequences Luhmann draws from his theoretical point of view are surprisingly conventional. This theoretical figure reminds of the classical bourgeois exaction of a stable and continuous identity. It looks like a caricature of the self-reserved, self-seeking, differentiated protestantic individual, that comes to himself or herself by a self-description that is able to reconcile the differentiated perspectives of modern multi-inclusion in terms of a higher level of subjective unity. Or this individual suffers under the fundamental implacability of differences and witnesses this experience as a source of unity and identity.

I repeat: The profit of this theoretical figure is its sensibility towards the structural societal background of the emergence of semantics of individuality and individualization. The concept *exclusion individuality* is able to emphasize the functionality of individualization processes for the modern form of multi-inclusion. But it has two critical points. *First*, it is only able to describe individualization processes as a result of the shift in societal structure toward functional differentiation. But this does not mean that individuals are compelled to describe themselves as individuals in the sense of an academic or intellectual prejudice. The structural individualization process also brings forth forms of self-description without *individualized* cultural figures in the sense of the self-stylization as *exclusion individuality*. Structural conditions and possible semantics of self-description have to be strictly differentiated (Nassehi, 2003, p. 106). The *second* critical point is that the theory of exclusion individuality does not have in mind the meaning of the function systems themselves for individualization processes and individualized self-descriptions. Actually, systems theory makes a peculiar distinction between the sphere of the emergence of an individual will and a self-referential form of self-description and the social structure of society. Even if a theory like this falls into the trap, to continue the narrative of the classical bourgeois form of individuality, we have to consider the power of that narrative that locates the individual will besides and beyond society.

I want to emphasize in the following that what we call *individualization* is not a process of externalization of individual life-forms from the function systems. In contrast to this, I feel that all function systems create their own programs of *individualized* and *individualizing* forms of communication.

## Inclusion individuality

Usually the importance of the function systems, that is, politics, of economic structures, of educational chances or legal rules, for individualization processes has been discussed with regard to the degree of freedom of individual decision making. I have the impression that this question itself is part of the public individualization semantics that do not bear in mind the fundamental sociological question of how deciding communications, bifurcations in social processes, the exaction of freedom in a world with limited resources can become attributed to individuals. I want to show that these attributions and the social construction of individuality and subjective autonomy are central elements of modern function systems and their programs. Given that, individuality, then, is not only a result of self-reflection and self-description in the *exclusion area* of the function systems, but also the result of individualizing programs of the function systems. In this respect, individualization has to be conceived both as *exclusion* and *inclusion*.

Another point is that function systems should not be conceptualized as spaces one could enter or not, similar to rooms with doors or houses with an entry area. The clue of the theory of autopoietic social systems is that function systems are not spaces like this, but only *spaces of communication*, meaning that function systems have emerged around *symbolically generalized communication media*. The most successful and most relevant communication in modern society is played out through these communication media: that is, truth, love, money, power, law, belief or art (Luhmann, 1995, p. 161). These media's function is to increase the chance of connectivity of communication in a world in which exactly this is improbable. These communication media establish expectations and structural stability by using media that are stable over time. And they are not restricted to special areas of modern society. Their only limitation is their communicative connectivity, and the success of this differentiation form can be explained with the ability of these functional communication forms to penetrate society as a whole – even if this can always be a *particular* wholeness only. I cannot see that these central, structurally mediated communication forms should not be relevant to the specific modern form of individuality – and with regard to systems theory this can only mean: to *communicative forms of individuality*.

I want to briefly give some hints about this concerning function systems' construction of individualized communication. Take, for example, the modern *economy*. As one of the usual suspects to be an obstacle to successful *individualization*, the economy uses communicative forms that bring forth the individual as a decider. The money media is based on the fundamental difference of possessing/not possessing. And possession is strictly bounded to an attributable address with a name and with supposed decision resources. The whole economic system is based on *individual actors* who are non-transparent for each other and who are able to continue their own strategies of handling money. This is true both for investors and for consumers, both for owners and those who try to become owners. The central program of the economic system – the narrative of the *homo oeconomicus* – bets on individual deciders who continue the dynamics of markets and prices, and, for example, the labor market, is nothing else but a market of individual demanders dealing with short resources. The contemporary economic system is not only dependent on *individualized* deciders, consumers or investors. It also brings forth this form of individuality by using its own communication media and its own code. This form of *individuality* doubtlessly is made by *inclusion*.

A similar mechanism can be observed in the *legal system*. The emergence of a special media for legal rights also presupposes attributable individual addresses that can be made responsible for their decisions and actions. The semantic forms of legal equality, responsibility, criminal discretion, legal claims to specific rights, general human rights, and contractual capacity all refer on the autonomy of individuals. Beyond that the modern legal system is strictly connected with the economical program of individual possession. It follows that the legal construction of individual responsibility is one of the central individualizing structural components of modern society. This figure of responsibility as an individualizing technique has widespread impacts on the production of subjectivity beyond legal affairs in a narrower sense (see also Gergen, Chapter 8 in this volume). Corresponding to my economics example, the legal system, hence, does not presuppose but brings forth communicative forms that constitute a special kind of individuality as personal uniqueness.

My next example is the *political system*. The modern political system has the societal function to produce collectively binding decisions and to enforce their authority. The political system has to produce collectivities and the visibility of a collective space that becomes aware of itself. Although the political system produces collectivities, modern political programs deal with the problem of establishing a collective link of individuals. Therefore, both the idea of the modern nation and the idea of democracy are based on individual aspirations that have to become bound. Beyond that, contemporary forms of policies and politics convert societal or collective problems into individual problems. One should not underestimate the central meaning of individual entitlement to payments, services, and support in modern welfare states for the emergence of individualistic self-descriptions. Although these entitlements are often looked upon as mechanisms that hinder individual responsibility, they produce a rather stable *individualized* view of one's own life.

*Education* and *religion* are perhaps the function systems with the most immediate effects on the constitution of individuality. Whereas education produces individual perspectives on the life course and the idea of an inner development of the person, the role of religion for the Western process of modernization is well known. The Protestant, but at last oecumenical idea of a personal relationship between the faithful and God was one of the central Western semantic figures establishing different kinds of self-observation and self-description. Although the awareness of sin and the redeeming function of self-inspection have decreased over the last generations, in contemporary society religion still has an individualizing effect. Perhaps the only *wholeness* religion can treat nowadays is the communicatively constructed *wholeness* of individual lives as an *inclusive* reflection on *exclusion individuality*.

Ultimately *medicine* is the function system with the most effective impact on 'Practices of the Self,' which reign over the modern subject, or better: which give modern subjects reasons to reign over themselves (Lupton, 1995). Health, or a healthy life, requires techniques of self-observation which use strong social imperatives but which have to be operated by a self that discovers his or her individuality within the practice of a healthy lifestyle. Medical requirements can be executed with highly moral claims. Inclusion into the medical system thus produces an inclusion individuality of the healthy subject – or its contradiction: the failed, ill, unhealthy subject which can no longer hark back to a relieving destiny, but which has to assume responsibility for its situation.

Similar individualizing effects could be shown in the *mass media system* with its preference to individualize its topics, or in the system of *social help*, where the effects of societal problems become solved as individual problems, or in the *scientific function system* where despite the classical concept of a truth that is independent from its observation, the observing researcher becomes treated as a particular, sometimes ingenious, individual, or at last in *modern arts* that presuppose unique individuals with unique inner worlds both on the side of the artistic producer and on the side of the consumer of arts. What I want to emphasize is that the function systems themselves produce emergent routines and semantics of constructing individual addresses for their communication. In that sense *exclusion individuality*, of course, is an ecological condition for the reproduction of the function systems. But this ecological condition has been established by the function systems themselves and their individualizing forms of *inclusion*. The codes and communication media and the programs of the function systems create accommodating conditions for coupling radical societal differentiation with the individual's need to emerge as a reflexive, continuous operator: the one to whom responsibility for making decisions is attributed. Perhaps this idea can help to disenchant the individualization narrative and to keep a distance from the bourgeois-academic or avant-garde presentation of the individualization debate. And I fear that the concept of *exclusion individuality* seems to be quite fascinated by that form of individuality.

## Some Conclusions

At the end of my chapter I want to come to some short conclusions. I wanted to show that the idea of deconstructing subjectivity, free will, or the well-known derivates often criticizes what has never existed. The semantics of subjectivity has emerged as a reaction to a special kind of both under-determination and over-determination of the individual in bourgeois society. So the philosophical idea of the subject also used this twofold form: under-determination in terms of freedom and authority, over-determination referring to the source of subjectivity (reason, spirit, etc.). As I have suggested, sociology deals with similar problems and uses the concept of the actor both to deconstruct the classical theory of subjectivity and to save it. Thus, great parts of sociology *believe* in the actor, not mentioning that the actor himself/herself is a point of attribution – and so sociology treats as a solution what has to be its proper problem: how the actor becomes an actor and why modern society counts on the free will of individuals (Nassehi, 2006, pp. 69–164).

By way of example, I then presented a sociological alternative with Niklas Luhmann's theory of inclusion – a theory built upon categories which are far from the classical tradition, but with its concept of *exclusion individuality* also falling into the trap of the classical narrative. My examples for *inclusion individuality* were then able to show that the individual becomes formed by the function systems and not outside of them. Function systems *use* the free will of individuals for their own stability.

Does the bourgeois constellation already exist? If it is right that free will can only be thought if it becomes limited by reason, by reasonable authorities, or by subordinating under a generality, then we can doubt if free will can be thought of anymore as a matter of fact. The connectedness of free will and authorities or of individual responsibility and good reasons is a special form of a traditional modernity which was able to domesticate the notion of free will – not to abolish it but to enforce it. In a society without those authorities, with the decline of professional expertise, and with the self-deconstruction of reason, free will is exposed to a peculiar fate. Free will has become a grantable power, and therefore it escapes more and more. In a society without stable standards of the right things, of reasonable reasons, and of 'good taste,' there is nothing a free will could achieve. I do not bemoan that – but I also do not praise it. Perhaps the idea of free will will come back to society, when the promises of genetic production of human beings and the idea of a material foundation of individual aspirations come true. In such a world free will would have a satisfactory antagonist.

Probably brain research could be able to re-establish free will. The discussion about Benjamin Libet's (2004) experimental research about the temporal lag between brain activities and consciousness, for example, is not interesting concerning the question of whether free will really exists or if there is a scope for – well, for whom, for the subject, for the brain, for the mind, for the person, for the individual, or only for the grammatical me? What we can learn from Libet's experiments and from the successors (e.g. Haggard and Eimer, 1999) is that free will cannot be observed but can only be attributed. From a sociological point of view, the results of Libet's research are not as amazing as they seem at first sight. We are familiar with the idea that subjectivity as awareness of being the autonomous author of the own activities is the result of a non-observable lag. The attribution to an autonomous subject is a result of historical experiences with endangered autonomy; the attribution to a responsible individual is a result of social practices which compel us not to feel compelled. That most of our decisions are pre-decided by social practices, structures of expectation, and by the selectivity of social order remains mostly unreflected – and this non-awareness itself is part of these social practices. The time lag between such pre-decisions and our awareness is probably much greater than the time lag between brain activities and consciousness, by all means more than the ominous 500 ms the brain is faster than us. The subject as an attribution address nowadays seems to be wedged between the structure of social practices and brain practices. Probably we can only assert a free will in times in which the self-description routines of autonomous agency is under fire – in early modernity under the fire of social hierarchies, endangered participation, compelling traditions, and a self-imposed state of tutelage (Kant); nowadays under the fire of a materialistic attribution to brains, genes, and nature. The semantics of a free will and of subjectivity has been a successful semantic protest against tradition. And so a sociological inquiry can state that the person is an effect of communication, but free will probably will return as an eventual effect of brain research.

## Note

\* I want to thank Sabine Maasen, Barbara Sutter, and Daniel Lee for really helpful comments.

## References

Adorno, Th. W., *Aspects of Sociology*, trans. John Viertel (Boston, MA: Beacon Press, 1972).

Foucault, M., *The Birth of the Clinic: An Archeology of Medical Perception*, trans. A. M. Sheridan-Smith (London: Tavistock, 1973).

Foucault, M., *Discipline and Punish: The Birth of the Prison*, trans. A. M. Sheridan-Smith, (Harmondsworth: Penguin, 1977).

Haggard, P. and M. Eimer, 'On the Relation Between Brain Potentials and the Awareness of Voluntary Movements', *Experimental Brain Research*, CXXVI (1999): 128–33.

Hegel, G. W. F., *The Philosophy of Right* (Oxford: Clarendon Press, 1967).

Hegel, G. W. F., *The Phenomenology of Spirit*, trans. A. V. Miller (Oxford: Oxford University Press, 1979).

Husserl, E., *On the Phenomenology of the Consciousness of Internal Time, 1893–1917* (Dordrecht: Kluwer Academic, 1991).

Kant, I., *Critique of Pure Reason*, trans. Werner S. Pluhar and Patricia Kitcher (Indianapolis: Hackett, 1996).

Libet, B., *Mind Time: The Temporal Factor in Consciousness* (Cambridge, MA: Harvard University Press, 2004).

Luhmann, N., 'The Autopoiesis of Social Systems', in R. F. Geyer and J. van der Zouwen (eds), *Sociocybernetic Paradoxes: Observation, Control and Evolution of Self-Steering Systems* (London: Sage, 1986), pp. 172–92.

Luhmann, N., Gesellschaftsstruktur und Semantik. Studien zur Wissenssoziologie der modernen Gesellschaft, vol. 3 (Frankfurt/M.: Suhrkamp, 1989).

Luhmann, N., *Essays on Self-reference* (New York: Columbia University Press, 1990).

Luhmann, N., *Social Systems* (Stanford, CA: Stanford University Press, 1995).

Luhmann, N., *Observations on Modernity* (Stanford, CA: Stanford University Press, 1998).

Lupton, D., *The Imperative of Health: Public Health and the regulated Body* (London: Sage, 1995).

Mills, C. W., 'Situated Actions and Vocabularies of Motive', *American Sociological Review*, V (1940): 904–13.

Nassehi, A., *Differenzierungsfolgen. Beiträge zur Soziologie der Moderne* (Opladen: Westdeutscher Verlag, 1999).

Nassehi, A., *Geschlossenheit und Offenheit. Studien zur Theorie der modernen Gesellschaft* (Frankfurt/M.: Suhrkamp, 2003).

Nassehi, A., *Der soziologische Diskurs der Moderne* (Frankfurt/M.: Suhrkamp, 2006).

# Part III
# Self and Politics

## Introduction

Politics always implies an idea of the persons governed. Consequently, this section brings together analyses of political developments paying due heed to the relevance will is afforded in each of them. 'What is government itself, but the greatest of all reflections on human nature?' (Madison quoted in Barber, 1984/2003, p. 679) – with reference to James Madison, Benjamin Barber identifies it as a commonplace 'that particular understandings of political life are intimately associated with particular views of human nature' (1984/2003, p. 679). This 'commonplace,' however, is pivotal for the analyses of political rationalities. As political rationalities address the adequacy of political actions and authorities along principles and ideals, they have both an epistemic and a moral form. In this vein, the idiom of political rationalities proves to be not just mere rhetoric, but the performative articulation of reality's amenability to political deliberations (Rose and Miller, 1992, p. 177). In particular, the capacity of self-regulation by way of individual autonomy is a key issue: its specific conceptions vary between holding it as a prerequisite, a reason, or an effect of political deliberation, and it is around this issue that different political rationalities and governmental technologies can be discerned.

The self as a willing one is at the core of liberal political thought: assuming that individuals are rational maximizers of pre-political preferences, liberalism holds the liberty of the person as a primary value. This *a priori* sets a tight limit to state action. Any state action aiming at other objectives than the coexistence of freedoms is deemed illegitimate. The individual is free to do anything except to harm others; he or she is responsible for whatever action is pursued. Welfarism, however, focuses on the individual in two totally different views, the dangerous and the endangered individual, and relies on an idea of the social that is completely absent in liberal thought. Taking the social as 'a causal force with its own natural laws' (Cruikshank), problems are not to be reduced to a result of individual failures but are to be viewed from the interdependence between people. The social constraints and conditions of voluntary action thus come to the fore giving rise to a more complex account of what political action is about: on the one hand, the defense of society against dangerous individuals and, on the other hand, the defense of individuals against the risks of sociality (Donzelot, 1988).

Against such an idea of the social and its implications, Margaret Thatcher's often-quoted statement 'there is no such thing as society' (1987) signaled the restoration of the liberal idea of society as merely the sum of individuals solely responsible for their own well-being. Neoliberal politics thus is based on the responsibilization of individuals using free will as the basic criterion for attributing responsibility. In this understanding, responsibilization operates as a building block for creating selves that willingly refrain from endangering society.

Indeed, the rise of neoliberalism 'is accompanied by the proliferation of new technologies of government that "arise out of the shifting of responsibilities from governmental agencies and authorities to organizations without electoral accountability and responsibility (for example, the 'privatization' of 'public' utilities, the civil service, prisons, insurance and security)" (Isin and Wood). In fact, it can be said that the task of government today is no longer engaged in traditional planning, but is more involved in enabling, inspiring, and assisting citizens to take responsibility for social problems in their communities, and formulating appropriate orientations and rationalities for their actions' (Ilcan and Basok, 2004, p. 132). The idea of those governed is a decisive element inherent in such formulations.

In this vein, John Clarke, Janet Newman, and Louise Westmarland in Chapter 5 explore the role of conceptions of the consumer in the reform of public services in the UK. New Labour's articulation of the citizen-consumer started off as a critique of a 'rationing culture'. In a neoliberal style of thought, New Labour locates the consumer as a 'willing self': a subject capable of self-direction, but hitherto unreasonably constrained by state or regulatory conditions. Furthermore, New Labour's consumer discourse inverted the associations of market as an inequality-producing mechanism and state as an equality-producing mechanism: New Labour argued that the state and its public services have created inequality and that choice could be the remedy to produce equality, for example, along issues of race/ethnicity, gender, sexuality, age, and disability.

Clarke and his colleagues employed questionnaires, interviews, and group discussions to find out whether users of public services viewed themselves as consumers of services. Their study showed that a significant number of people neither identified as consumers nor as citizens, but located themselves either in 'patient/user' or 'public/community' relationships. With regard to these findings, Clarke and colleagues challenge what, as they maintain, a number of studies of governmentality assume too readily: that discourses translate into practices and that discursively constituted subjections evoke the subjects they seek. As the interviewees showed skeptical, cynical, distanced, and reluctant responses when talking about themselves as being expected to be 'consumers,' Clarke and colleagues portray them as 'subjects of doubt' answering back to the discursively constituted subjections and ultimately proving to be rather unwilling selves.

In the guise of US neoconservatism, Barbara Cruikshank, however, observes a political conception that diagnoses neoliberal politics and its discursive preoccupation with willing selves as working far too well: whereas neoliberal policies rely upon the autonomy and economic rationality of the individual will to replace governmental functions, neoconservatives deem the willing self an effect of neoliberal governance that should be tamed by measures of remoralization. Market rationality without any state intervention, so goes neoconservative belief, cannot produce the moral ground on which it stands and therefore the neoliberal retreat of state action in favor of the market principle is a fatal undertaking. 'Neocons' claim authoritarian state intervention to be the appropriate remedy against a force produced by neoliberal policy and conceived of as dangerous: the free will. In Chapter 6 Cruikshank takes up the question of whether or not entirely new forms of power have emerged in the present political context, which she characterizes as 'neo-politics'. Her analysis concentrates upon neoconservative reforms of disciplinary power that take sexuality and family formation as their targets in the effort to revive the 'traditional family' and civil society. These reforms are revealed as a new configuration of power/knowledge in which the will is no longer useful as an instrument and effect of power. Rather, the will is treated as an obstacle to good government.

Mariana Valverde in Chapter 7 focuses on yet another transformation in governing practices and their conceptions of the persons governed. She describes this transformation as a movement from state-wide social planning to what she calls targeted governing. Governance of this type dispenses with the policing of cities or nations as a whole; instead, high-risk spaces, high-risk people, or risk factors are identified as targets of intervention. As city or state-wide social planning requires surveillance over a whole territory and population, this transformation might be deemed the death of the dream of the panopticon caused by practical and legal difficulties. Valverde, however, does not see targeted governance solely as a more modest proposal motivated by the recognition of the panopticon's failures. According to her observation, targeted governance is not only touted as more practical or more respectful of privacy than universal governance, but also as more desirable in normative terms. Now that the effort to govern society or the person as a whole has come to appear utopian or even dangerous, other operations are called for, operations aligning power with particular knowledge practices such as 'evidence-based' medicine.

By way of exploring the contemporary medical treatment of alcoholism, Valverde shows that instead of systematically treating the person identified as an alcoholic as a whole, drugs are used to selectively affect an isolated process in the brain in order to produce an adjustment. As the ideas of normalization and pathologization of the person give way to the adjustment and correction of brain disorders, identity-based governance may be waning. Nevertheless, a new identity regime and a new panopticon can be observed to be emerging on the basis of the identification of ever-new forms of 'targets' for governance.

Whereas the practices scrutinized by Valverde suggest an end of the person 'as we know it,' the arrangements analyzed by both Cruikshank and Clarke and his colleagues all adhere to concepts of personhood. In either case, however, the will proves to be a specific target of self-regulation – a target to be affected by medication, remoralizing measures, or the demand to choose.

## References

Barber, B. R., *Strong Democracy: Participatory Politics for a New Age* (Berkeley: University of California Press, 1984/2003).

Donzelot, J., 'The Promotion of The Social', *Economy and Society*, 17, 3 (1988): 395–427.

Ilcan, S. and Basok, T., 'Community Government: Voluntary Agencies, Social Justice, and the Responsibilization of Citizens', *Citizenship Studies*, 8, 2 (2004): 129–44.

Rose, N. and P. Miller, 'Political Power Beyond the State: Problematics of Government', *The British Journal of Sociology*, 41, 2 (1992): 173–205.

# 5
# Creating Citizen-Consumers? Public Service Reform and (Un)Willing Selves

*John Clarke, Janet Newman and Louise Westmarland*

This chapter explores the role of conceptions of the consumer in the reform of public services in the United Kingdom. In such reforms the consumer has embodied both a specific vision of modernity and a model of the agentic 'choice making' individual. We examine the way that the figure of the consumer has been enrolled into political and governmental discourses of reform and its problematic relationship to the figure of the citizen. We then consider responses from people who use public services: exploring their preferred forms of identification and conceptions of the relationships that are at stake in public services.[1] These responses indicate a degree of skeptical distance from governmental address and point to problems about the effectiveness of strategies of subjection. We conclude by considering the analytical and political significance of unwilling selves as dialogic subjects.

## Reinventing citizens as consumers

In recent attempts to reform public services in the United Kingdom, the figure of the consumer has played a starring role. Narratives of the citizen-as-consumer identified the rise of a consumer society or a consumer culture as driving the need for change in public services. Such terms mark a distinctive break between the past and future of public services:

> Many of our public services were established in the years just after the Second World War. Victory had required strong centralized institutions, and not surprisingly it was through centralized state direction that the immediate post-war Government chose to win the peace. This developed a strong sense of the value of public services in building a fair and prosperous society. The structures created in the 1940s may now require

change, but the values of equity and opportunity for all will be sustained. The challenges and demands on today's public services are very different from those post-war years. The rationing culture which survived after the war, in treating everyone the same, often overlooked individuals' different needs and aspirations ... Rising living standards, a more diverse society and a steadily stronger consumer culture have ... brought expectations of greater choice, responsiveness, accessibility and flexibility.

(Office of Public Service Reform, 2002, p. 8)

This notion of a consumer culture/society involves a particular view of the practice of consumption and the identity of the consumer that are taken to mark a distinctive phase of modernity. Although formally consumption refers to the practice of making use of, or even using up, objects, here consumption is equated with market exchange mediated by the cash-nexus. In the process, other practices and locations of consumption are subsumed in the generalization of the exchange model (Clarke, 1991, ch. 4). Similarly, the consumer becomes construed as a person (an individual) who forms choices and realizes them through money (or functional substitutes, such as theft or credit/debt). The defining feature of the consumer is thus the act of purchase: commodified goods, services or experiences are the means to consummating needs, wants or desires. It is this historically and culturally specific understanding of consumption and the consumer that provides the reference point and the discursive resources for imagining citizens as consumers of public services (on the variations of the consumer, see Maclachlan and Trentmann, 2004; and Trentmann, 2006). This individuating conception of the empowering or liberating character of purchase is a core element of what Thomas Frank has called 'market populism' (2001) and is intimately entwined with the emergence of neoliberalism (whether this is understood as an ideology or a mode of governmentality, see Larner, 2000).

Neoliberalism locates the consumer as a 'willing self': a subject capable of self-direction who has, hitherto, been unreasonably constrained by state or regulatory conditions. The consumer thus embodies 'private' rather than 'public' authority (Hansen and Salskov-Iversen, forthcoming). The consumer is thus threaded into the neoliberal imaginary as a critical figure for constructing the antagonism between the state and the market as forms of social coordination. 'Liberating' the consumer provides one critical imperative for dissolving the state/market distinction by enlarging the reach of the market. In the United Kingdom, the Thatcher-led Conservative Party that came to power in 1979 inaugurated programs of marketization and privatization of public services, including a central discursive role for ideas of 'choice'. Although there were other views of 'consuming' public services before then, the trajectory of the contemporary figure of the citizen-consumer took off from this political-cultural conjuncture (and was echoed in other Anglophone states). Choice – and its capacity to articulate the contrast between the active consumer and the passive citizen – was a key feature of Thatcherite anti-statism and anti-welfarism. The transnational New Right articulated this as the difference between the virtues of the Market in contrast to the oppressive, inefficient, and monopolistic State (variously conceived of as bureaucracy, as hierarchy, as monopoly provider of public services, and as state socialist societies).

Public choice theory created a political and intellectual space for the articulation of the citizen-as-consumer. It provided an 'economic' critique of public bureaucracies (e.g., Niskanen, 1971; Dunleavy, 1991). Pointing to the perverse combination of absent market disciplines and the presence of incentives to 'empire building,' careerism, and an inwards focusing of organizational attention, public choice theory challenged claims about such bureaucracies being led or guided by a public service ethos. On the contrary, the approach suggested that public servants were just as self-interested and venal as everyone else – but were not inhibited in the pursuit of such self-interest by the challenges and constraints of market dynamics. In elaborating this view, public choice theory distinguished between Producer interests and Consumer interests – with bureaucratic monopolies being driven by Producer interest at the expense of the Consumer.

The consumer/choice link was a potent feature of several aspects of Thatcherism's remaking of the welfare state and public services in the United Kingdom during the 1980s. Most notable were the decisions to enable tenants to buy council houses in the Housing Act of 1980 and the creation of a 'quasi-market' to enable school 'choice' in the Education Reform Act of 1988 (see, for example, Forrest and Murie, 1991; Gewirtz, Ball and Bowe, 1995; Fergusson, 1998; Pryke, 1998). Choice was construed as the defining characteristic of the consumer's relation to public services and had a complex relationship to 'marketizing' processes (see, inter alia, Bartlett, Le Grand and Roberts, 1998; and Taylor-Gooby, 1998). The 'right to buy' in housing dissolved (partially) the distinction between the public and private sector to shift public resources (at a subsidized price) into private ownership. 'Choice' in schooling gave parents (the proxy consumers of education) the non-cash mediated right to express preferences about the school that they wished their children to attend. As a result, parents attempted to choose schools – and schools got to choose children (and their parents). Elsewhere, 'market stimulation' was intended to spend public money on creating a market of competing providers (for example in the field of domiciliary and residential care after the 1990 NHS and Community Care Act). Competition between providers (whether in an 'internal' or an 'open' market) was expected to drive down costs, improve efficiency and deliver better results to the users or consumers of services (Cutler and Waine, 1997, ch. 3).

The citizen-as-consumer was to take a further turn in the post-Thatcher landscape of British politics. The Conservative party in the 1990s turned to the 'consumer' interest as a way of realigning the relationship between the public, government and public services. In the abstract, Thatcherism had been profoundly antithetical to conceptions of the public and 'society,' inclining towards privatization either to the market or to the household. In contrast, the Major governments adopted a stance of improving the quality of public services. This commitment was embodied in The Citizen's Charter (launched in 1991) and the proliferation of other 'Charters' for a range of other public services. The Citizen's Charter (and its offspring) articulated a particular fusion of consumerism and managerialism in public service provision (Clarke, 1997; Pollitt, 1994).

## Public services and the consumer society

The return of a Labour Government in 1997 raised new questions about the future of public services. New Labour had strong continuities with Thatcherism and was profoundly shaped by Anglophone neoliberalism. But this underspecifies its character as a political project and program. Elsewhere we have argued that the transnational ambitions of neoliberalism need to be negotiated into specific national political-cultural formations – they cannot simply be imposed, at least in the context of Western nation-states (Clarke, 2004a). New Labour's public service discourse was marked by attempts to deal with potential sources of discordance, disagreement and opposition and suture them instead into the logics of neoliberalism. Such processes involve more than 'mere rhetoric': they involve a politics of articulation, drawing alternative political-cultural conceptions (and the social forces that are attached to them) into supporting, but subordinated, roles in the new project (this is discussed at greater length in Clarke et al., 2007).

The starting point for New Labour's articulation of the citizen-consumer was its critique of the 'old' formation of public services – and its anachronistic character in the 'modern world' of consumer culture. As the earlier quotation from the Office for Public Service Reform indicated, the New Labour project juxtaposed old and new in the distinction between a 'rationing culture' and a 'consumer culture'. It regularly recycled this distinction in the view of old state monopolies as being built on a 'one size fits all' model in contrast to the diversity of wants, needs, and desires in the modern world:

> Since every person has individual requirements, their rights will not be met simply by providing a 'one size fits all' service. The public expects diversity of provision as well as national standards.
>
> (Office of Public Service Reform, 2002, p. 13)

> ... we must respond to the individual's aspirations and needs, and we must reflect the desire of the individual to have more control over their lives. We must recognise that the one size fits all model that was relevant to an old industrial age will neither satisfy individual needs or meet the country's requirements in the years to come.
>
> (Blair, 2003a, p. 17)

> Thirty years ago the one size fits all approach of the 1940s was still in the ascendant. Public services were monolithic. The public were supposed to be truly grateful for what they were about to receive. People had little say and precious little choice. Today we live in a quite different world. We live in a consumer age. People demand services tailor-made to their individual needs. Ours is the informed and inquiring society. People expect choice and demand quality.
>
> (Milburn, 2002)

The figure of the consumer thus embodied the effects of major social changes, to which the 'old' model of public services was ill-adapted. But New Labour had to address social and political expectations that had not been met by Thatcherite programs of privatization, marketization, and residualization of public services (Newman, 2001). In the public as a whole, among public service workers and among party members, there has been a consistent view that public services are necessary and that they need to be improved (not least because of the effect of 18 years of Conservative degradation). In this field of expectation, we can see (at least) three key issues that New Labour's commitment to reform and modernization has to engage with, and incorporate. The first was the question of what political principles or values should drive these reforms (especially in the light of concerns about 'privatization by stealth' or the abandonment of 'Old Labour' commitments). What emerged was a typical

Third Way distinction between persistent values and changing means of enacting them in the 'modern world':

> The values of progressive politics – solidarity, justice for all – have never been more relevant; and their application never more in need of modernisation… At home, it means taking the great progressive 1945 settlement and reforming it around the needs of the individual as consumer and citizen for the 21st century.
>
> (Blair, 2002)

The second key issue was that of equality. The political juxtaposition of market and state in twentieth-century social democratic discourse involved a contrast between inequality-producing mechanisms (the market) and equality-producing (or inequality-remedying) mechanisms (the state). Reforming public services around principles that derive from contemporary forms of market society (the consumer, choice, etc.) needed to be negotiated against this view of inequality. This was done in two ways: first, New Labour argued that the state – in its public services – has created inequality; second, they claimed that choice could itself be a means of producing equality.

> To those on the left who defend the status quo on public services defend a model that is one of entrenched inequality. I repeat: the system we inherited was inequitable. It was a two-tier system. Our supposedly uniform public services were deeply unequal as league and performance tables in the NHS and schools have graphically exposed … The affluent and well educated … had the choice to buy their way out of failing or inadequate provision – a situation the Tories 'opting out' reforms of the 1980s encouraged. It was a choice for the few, not for the many.
>
> (Blair, 2003b)

A choice for the few, not the many, emerged as a significant anchoring point for the consumerist approach to reform. Indeed, critics of choice were challenged for their 'elitism': wanting to reserve the privileges of choice for the few. New Labour's consumer discourse also picked up on a range of challenges to public services around 'equality' issues – around race/ethnicity; gender, sexuality, age, and disability. From these (to continue the passage quoted earlier), New Labour articulated a need to make services responsive to diversity:

> Since every person has individual requirements, their rights will not be met simply by providing a 'one size fits all' service. The public expects diversity of provision as well as national standards. Government too

wants such standards, but not at the expense of innovation and excellence. So these goals must be complementary, and support each other in practice.

(OPSR, 2002, p. 13)

But where social movements drew attention to, and challenged, the relationship between patterns of difference and structures of inequality, the Consumer discourse treated diversity as an individual fact, as the earlier quotation from the Office of Public Service Reform indicated: 'The rationing culture which survived after the war, in treating everyone the same, often overlooked individuals' different needs and aspirations' (OPSR, 2002, p. 8). In this formulation, differences are not inequality-related (or generating). Rather, differences exist as individual characteristics or aspirations to which services should be more responsive (see Lewis, 2003, on conceptions of difference and diversity in social policy). Structural conditions that generated Equal Opportunities conflicts over forms of 'second class citizenship' are dissolved into a field of individualized idiosyncrasy.

Finally, the figure of the Consumer owed much to the neoliberal critique expressed in Public Choice theory. New Labour's appropriation of the figure of the Consumer borrowed this antithetical view of Producer and Consumer interests (and the role of government as the People's Champions against the Producer interest, see Clarke, 1997).

Public services... have to be refocused around the needs of patients, the pupils, the passengers and the general public rather than those who provide the services.

(Blair, 2002, p. 8)

One key means for breaking the hold of the Providers was to introduce 'contestability' – enabling and encouraging competing providers alongside (or even instead of) the public sector 'monopolies':

Our aim is to open up the system – to end the one-size-fits-all model of public service, which too often meant one supplier fits all, with little diversity, irrespective of how good new suppliers – from elsewhere in the public sector, and from the voluntary and private sectors – might be.

(Blair, 2003b)

The figure of the Consumer has been central to the New Labour discourse of public service reform. Other terms did not simply disappear: the figures of citizens, communities, the public, users of services continued to appear. So, too, did more service-specific terms such as patients, passengers, pupils, and parents, when health, public transport, and education are being discussed. Nevertheless, these were increasingly subordinated to the idea of the consumer (and/or customer), a process described by Hall (2003) as 'transformism'.

## Transforming citizens: the consumer as a neoliberal archetype?

For both political economy and governmentality approaches to neoliberalism, the consumer is a central figure (Clarke, 2006; Clarke and Newman, forthcoming). It is a core image for the neoliberal claim about the nature of the world and how it must be. For example, Nikolas Rose, discussing advanced liberal governmentality, argues that:

> In this new field, the citizen is to become a consumer, and his or her activity is to be understood in terms of the activation of the rights of the consumer in the marketplace.
>
> Consider, for example, the transformations in the relations of experts and clients. Whilst social rule was characterized by discretionary authority, advanced liberal rule is characterized by the politics of the contract, in which the subject of the contract is not a patient or a case but a customer or consumer ... Of course, these contracts are of many different types. Few are like the contracts between buyer and seller in the market. But, in their different ways, they shift the power relations inscribed in relations of expertise. This is especially so when they are accompanied by new methods of regulation and control such as audit and evaluation ... The politics of the contract becomes central to contests between political strategies concerning the 'reform of welfare', and to strategies of user demand and user resistance to professional powers.
>
> (Rose, 1999, pp. 164–5)

The shift from citizen to consumer seems to embody a set of much wider distinctions: for example, from the state to the market; from rights to contracts; from the public to the private; from collectivism to individualism; from social democratic welfarism to neoliberalism; or from 'government from a social point of view' to advanced liberal rule. This list of distinctions demonstrates just how deeply the shift from citizen to consumer is understood as emblematic of neoliberalism. However, such a tidy list of binary distinctions might also make us think twice. Such second thoughts about the reliability and usefulness of the citizen/consumer distinction are reinforced by some of the results of our work on the construction of the citizen/consumer. In what follows, we explore these reservations about the shift from citizen to consumer in three different ways:

(i)   The relationship between the governmental project of constructing citizens as consumers and the political projects in which it has been embedded;
(ii)  The problematic relationships between identifications, relationships and practices; and
(iii) the problem of subjects who appear as unwilling selves.

The binary distinctions that we have sketched above tend to underestimate the political-cultural work that has been, and remains, needed to make neoliberalism possible: to make it look imaginable, plausible, necessary, and inevitable. This is at least a question about what work is needed to clear the ground of other orientations, other understandings, and other imaginaries so that neoliberalism can flourish. Neoliberalism does not enter a vacant or evacuated terrain: rather, it faces the challenge of displacing, co-opting or subordinating competing conceptions of the world. At the least, it needs to occupy this landscape in such a way that political and social subjects are compliant with a neoliberal sense of purpose and direction. For us, this raises questions about how the neoliberal project has been connected to, and voiced through, other politics, other discourses, or other rhetorics. New Labour did not simply announce that the consumer is the only possible project. As we have seen, it narrated the consumer as addressing and settling a whole set of other political, moral, and social problems. So New Labour made the consumer engage with questions of equality and reduced them to questions of equity. It engaged with a politics of difference, while reworking it into an issue about the infinite variety of individual difference. New Labour's consumerist orientation took up a variety of struggles around forms of power and forms of domination in public institutions, particularly those that challenged their organizational and occupational practices and their discriminatory exercise of power and authority, and condensed them into the demand for 'choice for the many'.

Both political economy and governmentality conceptions of neoliberalism risk short-circuiting these forms of political work (though for different reasons). The temptation is to treat neoliberalism as a relatively coherent, universalizing, or globalizing project. 'Politics' in its narrow institutional sense appears as either irrelevant, or as the rhetoric through which class interests are simultaneously spoken and concealed. In Foucauldian approaches in particular, the intentional widening of the concept of politics to the entire field of power/knowledge formations has tended to displace attention from the narrower institutionalist forms. However, we continue to see both national formations (albeit transnationally constituted national formations) and the practices of institutional politics as central to understanding both the logics and limits of neoliberalism (Clarke, 2004a and b; see also Kingfisher, 2002, and Sharma and Gupta, 2006). The process of articulating neoliberalism is subject to significant national political-cultural variation and requires the work of governments to produce some of the critical conditions for the 'governmental project'. At a minimum, politics – in its institutionalist sense – mediates the possibilities for any neoliberal governmental project.

It is in this sense that we want to insist on treating neoliberalism as both a governmental project and a political project. As a governmental project it requires the rearrangement, remaking, or reinvention of the apparatuses, policies, and practices of governing peoples (Newman, 2005). The simplifying binary view of the shifts from citizen to consumer, state to market, and public to private underestimates the amount of governmental work that is needed to institutionalize neoliberalism in these assemblages of policies and practices. It is not just, as the policy literature sometimes says, a matter of an 'implementation gap'. Rather, we need to think of the apparatuses, the occupations, the organizations and the practices as already multiply contested. Conservative and critical forms of professionalism, varieties of managerialism, radicalized orientations in social work, health, and education – have all left their traces on the apparatuses of public provision. And so the accomplishment of a governmental project involves transforming the institutions themselves: reconstructing institutional forms, organizational designs, and occupational cultures.

One might view the long, 30-year history of public service reform in Britain as a succession of strategies of institutional reform in which the introduction of internal markets coexists with 'contracting out' and privatization, as well as with the introduction of new modes of management, which attempt to displace and subordinate professional practices (Clarke and Newman, 1997). The deconstruction of different sorts of occupational formations through attempts to de-unionize, de-professionalize, and re-professionalize around new criteria combine in a series of strategies to remake the institutional formation of government (Clarke and Newman, 2005). Despite claims about transformations, this looks like a slow, uneven and incomplete process. Indeed, the constant turmoil of innovation in these fields suggests just how recalcitrant and reluctant to move these institutional formations have proved.

There is more that could be said about these 'institutional' problems and their implications for a governmental project centered on producing the citizen-as-consumer. Here, though, we turn to some of the findings from a recent research project on 'creating citizen-consumers' in which we looked at three public services: health care, social care, and policing (Clarke et al., 2007). Using a mix of questionnaires, interviews, and group discussions, we explored a simple question with people: who do you think you are when you use public services? We were interested in whether people had a conception, or an identification, of themselves as consumers of services. Quite simply, hardly anyone identified themselves as a consumer or a customer in relation to public services, as Table 1 indicates.

In the interviews and group discussions people reasoned eloquently about why they are not consumers in relation to public services. In the following extract one health service user explores the complex field of identifications, and the relationships and orientations that they imply:

> With the health service as a national health service, it's more than, I feel it's more than just the services that you consume. I mean I am concerned with it more on the whole than just being consumers. So even if I wasn't attending the hospital or seeing my GP regularly, OK I'd still register with a GP, so from that point yes I would be a consumer, but it's not ... If I was 100% healthy and not using, consuming the services, I would still feel a relationship to the health service because I pay for it, it is not Tony Blair's or whoever's money, it's our money, we paid for it, it's the nation's, the national health service. And I do consider that when I cast my vote. So even if I wasn't actually in need of the service it still does affect me and I still would consider that at election time. So I feel it's more than just a direct consumer because you are paying for a national service for everyone's benefit. Whether you

*Table 1*   User identifications

|  | **Number** | **Answers %** | **People %** |
| --- | --- | --- | --- |
| consumer | 4 | 2.2 | 3.8 |
| customer | 8 | 4.4 | 7.5 |
| patient | 34 | 18.7 | 32.0 |
| service user | 35 | 19.3 | 33.0 |
| citizen | 19 | 10.5 | 17.9 |
| member of the public | 39 | 21.5 | 37.0 |
| member of local community | 42 | 23.2 | 39.6 |
| *Respondents could choose one or two identifications* | *N = 181* |  | *N = 106* |

actually need to consume that service or not, is not the primary consideration. So it's wider than just being considered a consumer, I feel.

(Newtown health user 3)

This statement maps a complex set of relationships and orientations that are at stake in one public service (the NHS). The idea that 'it's not like shopping' is one recurrent phrase that people used to denote the distinctiveness of public services (Clarke, forthcoming). People understand that the figure of the consumer references the experiences and practices of shopping and observe that their relationships to public services are never like that. Furthermore, when people use public health care, social care, or policing, they typically engage with them in a situation of distress. People use them to try to remedy a condition in which they do not wish to be: illness, vulnerability, being victimized. These are not freely chosen moments nor are they moments when people wish to be in the distant 'transactional' mode that they associate with being a consumer:

I don't like 'customer' really, because it implies a paying relationship on a sort of take it or leave it basis – more like going into a shop and seeing what's available and choosing something. I don't think it's quite like that …

(Newtown health user 1)

But while hardly anyone identified themselves as consumers, not many more identify as citizens (see Table 1 above). The term seems not to have the particular density that people want to talk about when discussing their relationship to public services. It seems somehow too abstract, too 'political,' perhaps, and not relational enough. Instead, we found two main ways of talking about public services. One is in very service-specific terms: most people talking about their relationship with health care stress the identification of being a patient. They talk in extremely complex ways about what it means to be a patient, but nevertheless understand 'patient' to define the particularity of their relationship to health care in the process of using or consuming it. They may at the same time see themselves as many other things: a taxpayer, a voter, a member of a consultative body, and so on. But 'patient' seems to best describe the location and the relationship that they find significant (Clarke and Newman, forthcoming):

'Patient' is the traditional term and I think it is still appropriate. The *NHS is a service* to users (in the local community). I know 'consumer' and 'customer' imply choice and that is what we are supposed to want. I would consider it an acceptable achievement if everyone could have what was best in the matter of treatment as of right. There are certain cost considerations but that is another issue. 'Choice' may be a political ploy to take our eye off the ball and confuse us as to what really matters. Choice sounds a good thing – but is it? Sorry, this is one of my hobby horses!

(Newtown health user questionnaire 23: patient and service user: emphasis in original)

We will return to this quotation later in the chapter to explore its reflexive complexity. But it is important to note that conceptions of membership formed the second key element in how our respondents identified themselves in their relationships with public services. Substantial numbers saw themselves as members of the public, or as members of local communities. The political cultural significance of 'membership' is a difficult conceptual issue, implying relationships of both inclusion and exclusion, but these two terms were consistently signaled as important ways of identifying the relationship with public services. The identification with being a 'member of a local community' may also be a way of people trying to capture the localness – the spatial specificity – of service provision. All of the services have local systems of provision, although the geographical arrangement of the 'local' varies, with few coterminous boundaries.

At this point, we want to stress the peculiarity or perverseness of these findings. They do not reflect a widespread shift towards a consumer or customer identification. The political and governmental project that New Labour represents has failed to embed itself in the identifications or self-conceptions of these users of public services. But neither do they seem to be grounded in an alternative social democratic or republican conception of citizenship. These two terms (citizen, consumer) are taken – in politics and in the social sciences – to be the binary framing of current changes. Yet people's own identifications seem to be located elsewhere – either in service-specific (patient/user) or 'public/community membership' relationships.

## Commitment and compliance: the constitution of subjects

There are a number of limitations to the study on which we report here. It treats subjection primarily as a matter of identification – do subjects sign up to, or occupy the dominant discursive figures? We think it is significant that they do not – both for New Labour claims about the social changes to which public services must adapt, and for governmentality arguments about the creation of enterprising selves or consumer identities in the process of 'economising the social'. We return to the analytical implications of people not identifying with these positions in the final section, but here we turn to the relationship between discourses and practices in the enactment of governmental projects. This poses the question of whether subjection is primarily a matter of commitment or compliance. Do subjects occur because they come to recognize themselves in the dominant discursive figures, or because the institutionalized relationships and practices require them to behave in certain ways?

For the construction of the citizen-consumer, this might mean that constructing fields of choice in which choices are obligatory positions people as if they are consumers. In such circumstances (parental choice in schooling; choice of hospital, and so on), people are required to act as consumers whether they understand themselves in such terms or not. That is to say, the 'conduct of conduct' may be accomplished by the regulating effects of a field of relationships and practices rather than through forms of identification. This may be what Margaret Thatcher called the TINA effect ('There Is No Alternative'): compliance is what is required rather than commitment. Such behavioral compliance may bring identification or subjective attachment in its wake, but it is not a necessary requirement. In short, people may behave like consumers, even if they do not think of themselves as consumers.

In one of our interviews (with a voluntary sector organization), a version of this 'compliance' model is explored, with the interviewee reflecting on both the popular distance from the consumer identity and the use of consumer-like practices to make demands on public services:

> I think what people want are good public services. I think they want good local deliverable public services. I don't think they want – I don't think they want to apply consumer principles to those. I don't think they want choice, I don't think they want competition and I don't think they want market forces. I think they want good, um, schools, good hospitals, good GPs ...
>
> I think people behave as though it's true ... I think people behave as if it's true so I think people when they're not happy with something, um, employ the techniques for dealing with it that they employ in a consumer, um, situation. So you know ... if they don't like what happened to little Johnny in school yesterday, they will go and challenge in the way that they might if their fridge broke down after three days. So I think actually people do employ those techniques ... I think that might be because they haven't been given any other skills in part and

because there are no other overt frameworks through which they necessarily understand they can do it … I mean I think the fact that
everybody … is now accountable means that everybody thinks they
have a right to challenge and there's lots of good that's come out of
that shift but I think the bad is that sometimes that's not the most
productive way of dealing with something and it's often not the most
appropriate. But it is consumerism. I mean that is what you do as a
consumer. You would be – the strength lies with you because you're
the purchaser and therefore that gives you the power. And I think
people have taken that and apply it to all, um, all areas of dispute.

(Newtown voluntary 01)

Here the suggestion is that the practices of being a consumer offer one,
and perhaps the only available, means of being assertive or demanding
in interactions with public services, even if people do not seek choice,
competition, or consumer principles. Being a consumer implies what
might be called 'transferable skills,' rather than transferable orientations,
principles, or identities.

We think that both these views of practices, rather than identifications,
significant for the installation of a neoliberal consumerist orientation
in public services, raise important issues. But they also open up further
problems about how we assess or evaluate such developments, particularly in terms of identifying their 'consumerist' character. Making choices
(particularly where people are compelled by state authority to make
choices) is not the same as being a consumer. Nor are we sure that being
assertive or demanding is the same as using consumer techniques. At
a behavioral level (leaving aside questions of identification and action),
we are not convinced that it would be possible to clearly distinguish the
practices of a choice-making or demanding consumer from an 'expert'
and 'co-producing' patient (committed to 'leading their medical team') or
from an assertive, rights-bearing citizen. There are multiple political, cultural, and personal routes to a sense of being 'entitled' (see, for example,
Cooper's analysis of conceptions of 'belonging,' 1998).

In part, these comments point to a problem about the conceptualization
of the consumer. Much of the writing about the consumer in relation to
public services (both positive and critical) treats the consumer as the
embodiment of economic models – rationalistic, calculating, and atomized.
For neoliberals and their allies, the consumer thus embodies the triumph of markets over repressive and constraining states, for critics it marks
the desocialization of the public realm. But both share a presumption
that consumers do indeed behave like consumers. The sociological

and cultural studies literature on consumers and consumption tends to undermine this economistic conception of the consumer – pointing to the varieties of rationality in play in consumption (and the valorization of 'irrationality' in dynamics of pleasure and desire), and stressing the diverse social dynamics (personal, familial, communal, subcultural, local as well as global) that shape the practices and choices of consumption. Such studies challenge the apparent coherence and unity of the consumer as s/he is imagined in economic terms (see, for example, Gabriel and Lang, 1995; Daunton and Hilton, 2001; Trentmann, 2006). Instead, they point to what Gabriel and Lang term the 'unmanageable consumer,' whose defining feature is precisely its unpredictability. Neither the idealized sovereign nor the despised dupe, the consumer reappears as a mobile and multiple subject. This returns us to questions about how to think about subjection in terms of subjects who are contingently willing and unwilling, and who are heterolingually dialogic, rather than trapped in a binary dynamic of acquiescence or resistance (Holland and Lave, 2001; Morris, 2006).

## Unwilling selves and dialogic subjects

In this final section, we turn to questions about the subjects of discursive practice. Elsewhere, we have argued that too many studies of governmentality too readily assume that discourses translate into practices, and that discursively constituted subjections evoke the subjects they seek (Clarke et al, 2007). Here, we focus on the second of these points. There are both empirical and analytical problems about assuming that the subjects summoned in and through discursive practice will come when called. As we have seen, our own study suggests that the identifications at work among the public fail to align with the consumer/customer orientation. In particular, people actively refuse the identification of being a consumer of public services – and the implied de-differentiation between public services and the market place. Nor do people grasp their relations to public services within the binary of citizen–consumer so central to contemporary public, political, and political science debates (Clarke, 2006; Clarke and Newman, forthcoming). Similarly, providers of public services expressed reservations about choice and consumerism as principles for the (re)organization of services, in part because of their dislocating effects on established occupational and organizational formations of knowledge and power.

The discussions of 'consumer' in our study are consistently marked by skepticism, cynicism, distance, and denial. These views are voiced by 'subjects of doubt' (Clarke, 2004c). Such subjects reflect upon the dominant discourse, its interpellations and the subject positions it offers. They reason about different sorts of identifications and the relationships they imply. They make choices about what terms evoke their desired personal and political subject positions. They embody key elements of what Holland and Lave (2001) call 'dialogism' – the Bahktin-derived conception of subjects who 'answer back'. The quotations above testify to people who know that they are being spoken to – and are reluctant to acquiesce or comply. One of the earlier quotations perfectly captures this dialogic reflexivity: 'I know "consumer" and "customer" imply choice and that is what we are supposed to want ... "choice" may be a political ploy to take our eye off the ball and confuse us as to what really matters.' There speaks a subject who hears the process of subjection ('that is what we are supposed to want'), recognizes its political-cultural character ('a political ploy'), and offers an alternative account of what we want: as a 'matter of right'. We think that such 'subjects of doubt' imply a form of analysis that pays attention to the fractures or disjunctures in the circulation of discourses – rather than assuming their success in recruiting or enrolling the subjects they seek (see also Marston, 2004).

It may be important to reflect that these are subjects who are already skeptical. Their skepticism means that they do not need the revelatory mode of academic analysis to demonstrate what they already know: that language and power are entwined; that words have consequences and implications; that the future is being constructed and is contested; that identifications matter. Neither tearing aside the veil of ideology nor uncovering discursive constitution seems adequate (either analytically or politically) to the way that such subjects live their relationship to social institutions and political practices. They are, of course, not outside discourse: it makes more sense to think of them as mobilizing multiple discourses to enable a space of skepticism about the dominant. They inhabit the world of 'common sense' in its Gramscian sense where 'traces' of heterogeneous philosophies, ideologies, discourses are layered up and may be put to use (e.g., 1971, pp. 324–5). This Gramscian view is, we think, different from more sociological conceptions that treat common sense as the forms of everyday knowledge always and already colonized by dominant understandings. In contrast, Gramsci was insistent about the multiplicity of common sense and about the implications of that multiplicity for the possibilities of political work and engagement. In particular, he stressed how – in political terms – common sense always contained elements of potential 'good sense,' rather than being merely regressive or reactionary. In the disjunctured and sometimes contradictory relationships between these different and divergent elements, 'perspectives' on the dominant may be opened up.

These questions about language, subjection, and skepticism point to a view of governing as a profoundly uneven and incomplete process in which subjects succumb, sign up, or comply but may also resist or prove recalcitrant and troublesome. In the process, attempted subjections are likely to be less than comprehensive and only temporarily settled. In short, we incline towards an approach that stresses a politics of articulation rather than a politics of subjection. We see a danger in studying the processes of subjection from the standpoint of the aspirations of the dominant point of view. The temptation is to see the world aligning with the plans, visions, or scripts of the dominant. Governmental projects like people to know their place. But people prove strangely – and unpredictably – reluctant to acquiesce. Starting from an unruly conception of the social as a field to be governed might enable a better view of the uneven and incomplete character of subordination and subjection. We may see the rich repertoire of ways in which people live their subordinations: the enthusiastic engagement, the calculating compliance, the grudging or foot-dragging recalcitrance, the practices of resistance or refusal, the elaboration of alternative possibilities. That implies looking for the ways in which people fail to 'know their place' – or sometimes remain overly attached to it, when authority would like them to move to a new place.

From this starting point, the social is a contested terrain in its own right, subjected to multiple and conflicting attempts at 'mapping' places, positions, relations, and differences (and all the inequalities that such differences may distribute). Some of those mappings are 'governmental' – the official classifications, distinctions, locations used to constitute populations. But the social is also a field of resources – identities, potential solidarities, languages, and voices – with which the subjected and subordinated may 'answer back' to the dominant and would-be hegemonic 'hailings' of authority. We do not mean to romanticize the social in drawing attention to its recalcitrance. The distance between people and intended subjections is not intrinsically progressive, nor even intrinsically political (in the sense of mobilizing collective action). However, as Chatterjee (2003) insists, the recalcitrant, difficult, and demanding existence of the 'governed' has profound political effects. It is possible, of course, that systems – economic, political, or governmental – may work without the complete subjection or subordination of their subjects. As we suggested earlier, grudging or calculated compliance may, indeed, be enough to make things work. Equally, passive – non-mobilized – dissent or skepticism may enable forms of political and governmental rule.

Nevertheless, the gaps between imagined subjection and lived identifications and attachments should alert us to the limits of plans and projects. In the process, we might also note the power and potential of both residual and emergent alternatives to the dominant – the elements and fragments of alternative futures (Williams, 1977: 121–3). That is why 'unwilling selves' and 'dialogic subjects' might be worth our attention.

## Notes

1   Creating Citizen-Consumers: Changing Relationships and Identifications was funded by the ESRC/AHRB Cultures of Consumption programme and ran from April 2003 to May 2005 (grant number: RES-143-25-0008). We studied three public services (health, policing, and social care) in two places (Newtown and Oldtown). We distributed 300 questionnaires (returns from 106 users and 168 staff = 46 per cent return rate). We conducted 24 interviews with managers; 23 with front-line staff; 10 with users and held 6 user focus groups. The project team was John Clarke, Janet Newman, Nick Smith, Elizabeth Vidler and Louise Westmarland, based in the Faculty of Social Sciences at The Open University, UK. More details can be found at: www.open.ac.uk/socialsciences/ citizenconsumers.

## References

Bartlett, W., J. Le Grand and J. Roberts (eds), *A Revolution in Social Policy: Quasi-Market Reforms in the 1990s* (Bristol: The Policy Press, 1998).

Blair, T., *Prime Minister Tony Blair's Speech at the Labour Party Conference in Blackpool* (Guardian Unlimited, 1 October 2002).

Blair, T., *Progress and Justice in the 21st Century*, The Inaugural Fabian Society Annual Lecture, 17 June (2003a). Available at: http://politics.guardian.co.uk/ speeches/story/0,11126,979507,00.html

Blair, T., *Where the Third Way Goes from Here*, Policy Network/Progressive Governance Conference (2003b). Available at: http://www.progressive-governance.net

Chatterjee, P., *The Politics of the Governed* (New York: Columbia University Press, 2003).

Clarke, J., *New Times and Old Enemies: Essays on Cultural Studies and America* (London: HarperCollins, 1991).

Clarke, J., 'Capturing the Customer: Consumerism and Social Welfare', *Self, Agency and Society*, vol. 1, 1 (1997): 55–73.

Clarke, J., 'Dissolving the Public Realm? The Logics and Limits of Neo-Liberalism.' *Journal of Social Policy*, 33, 1 (2004a): 27–48.

Clarke, J., *Changing Welfare, Changing States: New Directions in Social Policy* (London: Sage, 2004b).

Clarke, J., 'Subjects of Doubt.' Paper presented to Canadian Anthropological Society (CASCA) conference, University of Western Ontario, May (2004c).

Clarke, J. 'Consumerism and the Remaking of State-Citizen Relationships', in G. Marston and C. McDonald (eds), *Analysing Social Policy: A Governmental Approach* (Brighton: Edward Elgar Publishing, 2006).

Clarke, J., 'It's not like shopping: Relational Reasoning and Public Services', in M. Bevir and F. Trentmann (eds), *Governance, Citizens, and Consumers: Agency and Resistance in Contemporary Politics* (Basingstoke: Palgrave Macmillan, 2007).

Clarke, J. and J. Newman, *The Managerial State: Power, Politics and Ideology in the Remaking of Social Welfare* (London: Sage, 1997).

Clarke, J. and J. Newman, 'The rise of the Citizen-Consumer: Implications for Public Service Professionalism', Paper presented to C-TRIP seminar, King's College London, 19 October (2005).

Clarke, J. and J. Newman, 'What's in a name? New Labour Citizen-Consumers and the Remaking of Public Services', *Cultural Studies* (forthcoming).

Clarke, J., J. Newman, N. Smith, E. Vidler and L. Westmarland, *Creating Citizen-Consumers: Changing Publics and Changing Public Services* (London: Sage, 2007).

Cooper, D., *Governing out of Order* (London: Rivers Oram Press, 1998).

Cutler, T. and B. Waine, *Managing the Welfare State* (Oxford: Berg, 1997).

Daunton, M. and M. Hilton (eds), *The Politics of Consumption: Material Culture and Citizenship in Europe and America* (Oxford: Berg, 2001).

Dunleavy, P., *Democracy, Bureaucracy and Public Choice* (London: Harvester Wheatsheaf, 1991).

Fergusson, R., 'Choice, Selection and the Social Construction of Difference: Restructuring Schooling', in G. Hughes and G. Lewis (eds), *Unsettling Welfare: The Reconstruction of Social Policy* (London: Routledge/The Open University, 1998).

Forrest, R. and A. Murie, *Selling the Welfare State: The Privatisation of Welfare Housing* (London: Routledge, 1991).

Frank, T., *One Market Under God: Extreme Capitalism, Market Populism and the End of Economic Democracy* (New York: Anchor Books, 2001).

Gabriel, Y. and T. Lang, *The Unmanageable Consumer: Contemporary Consumption and its Fragmentations* (London: Sage, 1995).

Gewirtz, S. , S. Ball and R. Bowe, Markets, *Choice and Equity in Education* (Buckingham: Open University Press, 1995).

Gramsci, A., *Selections from the Prison Notebooks* (London: Lawrence & Wishart, 1971).

Hall, S., 'New Labour's Double Shuffle', *Soundings*, 24 (2003): 10–24.

Hansen, H.-K. and D. Salskov-Iversen (eds), *Critical Perspectives on Private Authority in Global Politics* (Basingstoke: Palgrave Macmillan, 2007).

Holland, D. and J. Lave, 'History in Person: An Introduction', in D. Holland and J. Lave (eds), *History in Person: Enduring Struggles, Contentious Practices, Intimate Identities* (Santa Fe: School of American Research; Oxford: James Currey, 2001).

Kingfisher, C. (ed.), *Western Welfare in Decline: Globalization and Women's Poverty* (Philadelphia: University of Pennsylvania Press, 2002).

Larner, W., 'Neo-liberalism: Policy, Ideology, Governmentality', *Studies in Political Economy*, 63 (2000): 5–25.

Lewis, G., '"Difference" and Social Policy', in N. Ellison and C. Pierson (eds), *Developments in British Social Policy* 2 (Basingstoke: Palgrave Macmillan, 2003).

Maclachlan, P. and F. Trentmann, 'Civilising Markets: Traditions of Consumer Politics in Twentieth-Century Britain, Japan and the United States', in M. Bevir and F. Trentmann (eds), *Markets in Historical Contexts* (Cambridge: Cambridge University Press, 2004).

Marston, G., *Social Policy and Discourse Analysis* (Aldershot: Ashgate, 2004).

Marston, G. and C. McDonald (eds), *Analysing Social Policy: A Governmental Approach* (Brighton: Edward Elgar, 2006).

Milburn, A., *Reforming Public Services: Reconciling Equity with Choice*, Fabian Society Health Policy Forum, Fabian Society, London, 2002.

Ministers of State for Department of Health, Local and Regional Government, and School Standards, *The Case for User Choice in Public Services*. A Joint Memorandum to the Public Administration Select Committee Inquiry into Choice, Voice and Public Services (London: House of Commons, 2004).

Morris, M., *Identity Anecdotes: Translation and Media Culture* (London: Sage, 2006).

Needham, C., *Citizen-Consumers: New Labour's Marketplace Democracy* (London: The Catalyst Forum, 2003).

Newman, J., *Modernising Governance: New Labour, Policy and Society* (London: Sage, 2001).

Newman, J. (ed.), *Remaking Governance: Peoples, politics and the public sphere* (Bristol: The Policy Press, 2005).

Niskanen, W. A., *Bureaucracy and Representative Government* (New York: Aldine-Atherton, 1971).

Office of Public Services Reform [OPSR], *Reforming our Services: Principles into Practice* (London: Office of Public Services Reform, 2002).

Pollitt, C., 'The Citizen's Charter: A Preliminary Analysis', *Public Money and Management*, April–June 1994: 9–14.

Pryke, M., 'Thinking Social Policy into Social Housing', in G. Hughes and G. Lewis (eds), *Unsettling Welfare: The Reconstruction of Social Policy* (London: Routledge/The Open University, 1998).

Rose, N., *Powers of Freedom* (Cambridge: Polity Press, 1999).

Sharma, A. and A. Gupta, 'Rethinking Theories of the State in an Age of Globalization', in A. Sharma and A. Gupta (eds), *The Anthropology of the State: A Reader* (Oxford: Blackwell Publishing, 2006).

Taylor-Gooby, P., *Choice and Public Policy: The Limits to Welfare Markets* (Basingstoke: Macmillan, 1998).

Trentmann, F. (ed.), *The Making of the Consumer: Knowledge, Power and Identity in the Modern World* (Oxford: Berg, 2006).

Williams, R., *Marxism and Literature* (Oxford: Oxford University Press, 1977).

# 6

# Neopolitics: Voluntary Action in the New Regime

*Barbara Cruikshank*

> As you know, no one is more of a continuist than I am: to rec-
> ognize a discontinuity is never anything more than to register a
> problem that needs to be solved.
>
> Michel Foucault (1980/2000, p. 226)

Michel Foucault's work on ethics and the care of the self is fertile ground
for thinking anew on the subject of willing selves. However, this chapter
aims to wed the 'practices of freedom' to *politics* on the basis of Foucault's
work, rather than turn as he did toward *ethics* (Foucault, 1991, p. 3). By
drawing out the singularity of reform and highlighting its political dimensions in Foucault's work on the prison and governmentality, I propose
that practices of freedom might be grounded in the ever-shifting grounds
of what counts as politics. The contingency of politics is captured by the
concept of 'neopolitics,' used here to indicate politics in a state of adaptation and change. In that contingency, I locate the possibility of practicing
politics and freedom in ways that can resist the shaping and instrumentalization of the willing self advanced by the forces of neoliberalism and
neoconservativism. In defense of that contingency, I combat the chorus of
contention alternately threatening and boasting that the present marks
the end of political contingency, such as Francis Fukuyama's declaration
of the end of history or the many voices announcing the advent of postmodernity. Rather than at the end, I situate the present in terms of its continuity with the past, both in order to claim politics as a practice of freedom
and to reclaim the contingency of politics.

## Perpetual reform?

Michel Foucault's work on the prison concludes that the perpetual reform
of the prison is the enduring and relentless principle of its longevity and

its productivity. The repetitive and monotonous critique of the prison for its failure to eliminate crime or rehabilitate the criminal, far from calling the prison itself into question, redoubles efforts to make the prison work. 'For a century and a half the prison had always been offered as its own remedy: the reactivation of the penitentiary techniques as the only means of overcoming their failure; the realization of the corrective project as the only method of overcoming the impossibility of implementing it' (Foucault, 1979, p. 268). The self-evidence of the prison as a solution to the problems of punishment was rather quickly established, after a long storm of reform, in the first half of the nineteenth century and remained so until the publication of *Discipline and Punish*.

However, for all its self-evidence, the prison was never stable or widely accepted as an institution of punishment. 'One should recall that the movement for reforming the prison, for controlling their functioning is not a recent phenomenon. It does not even seem to have originated in a recognition of failure. Prison "reform" is virtually contemporary with prison itself: it constitutes, as it were, its programme' (Foucault, 1979, p. 234). In other words, rather than born of legislation or the capitalist state, the prison was born of and is sustained by critique and debate over the right to punish and the proper techniques of punishment. Rather than accepted without thought, the persistent self-evidence of the prison is the result of constant agitation and critical thought.

Reform, then, is not a *response* to the criticism of corrective and disciplinary measures for failing to eradicate poverty, eliminate crime, moralize the masses, and so on. Rather, reform *is* the program of the prison, both in terms of reforming the individual convict and reforming society. 'The prison should not be seen as an inert institution, shaken at intervals by reform movements. The "theory of the prison" was its constant set of operational instructions rather than its incidental criticism – one of its conditions of functioning' (Foucault, 1979: 235). If reform and critique are conditions of the prison's function, then virtually any movement to reform the prison, however newfangled its techniques or programs posed as the solutions to the failures of the prison, should be understood as confirming the self-evidence of the prison as the proper way to punish. The perpetual reform of the prison is the condition of its longevity and, indirectly, the longevity of the liberal capitalist state.

Perpetual reform – or what Foucault came to call the governmentalization of the state – is the condition of the state's survival, a mode of fending off revolutionary struggle and absorbing political resistance despite the perpetual failure of the state to fully realize the security and welfare of the population. Yet, we are quick to ask, if reform is, in a more traditional vocabulary, the vehicle by which the state reproduces both itself and the willing consent of its people, then shouldn't a properly radical response be to eschew reform in order to rob the state of its capacity to restabilize itself, to open it up to revolutionary action? The question, and its opposition of revolution and reform, are as old as constitutional government and predate Foucault's publications. Given how readily the question is posed, it is clearly one that harbors enduring presumptions about political resistance. First, that resistance is *either* reformist *or* revolutionary *vis-à-vis* the state. Second, the tangibility of the state poses the ultimate question for critical thought and radical political action. The question imposes the state as the limit of our thought on the subjects of politics and political will. If, apparently, politics and political will can only be understood in relation to the state, if they are construed only as for or against, inside or outside, reproducing or transcending the state, then all possible political action is determined in advance.

However, another ordinary sense in which we discuss politics is that something is political (and not natural or given) insofar as it could be otherwise. Anything subject to change and to political will is 'political' and if it is out of our control, it may be a force to be reckoned with, but not 'political'. Rather than contradictory, these two notions of politics – as determined by the state, on the one side, and as covering everything that is not otherwise determined – combine to limit how far we might bring politics, in the sense of that which might be otherwise, to bear upon our conceptions of what is properly political. For example, unemployment may be out of control and we may debate whether or not to hold the state responsible for bringing the economy to heel. Setting the limits of state action, in this case, has everything to do with determining what is properly political and what is better left to market forces. Contemporary political thought, caught within the limit of the state, asks repeatedly, for example, if political will is conditioned by the state, can it ever be free? If politics is conditioned by the state, can it ever be democratic?

Foucault's conception of governmentality displaces the centrality of the state, as we shall see, and opens up the possibility that not only have we profoundly underestimated the political significance of reform as a vehicle for reshaping willing selves. We also underestimate the radical potential of reform as a vehicle for transforming what counts as political – in the sense of alterable or actionable – and thus the opportunities to create new grounds for the practice of freedom.

Foucault's essay on governmentality offers some insight into how that limit might be overcome: 'This governmentalization of the state is a singularly paradoxical phenomenon: if in fact the problems of governmentality and the techniques of government have become the only political issue, the only real space for political struggle and contestation, this is because the governmentalization of the state is, at the same time, what has permitted the state to survive' (1978/2000, p. 221). The 'paradoxical phenomenon' is a political formation – the liberal state – that relies upon and can be understood only on the basis of the tactics of government that comprise it and thus the state has no identity or solidity of its own. Yet, paradoxically, the reform and further governmentalization of the state is the means by which political conflict is captured by or contained within the state: 'the tactics of government that make possible the continual definition and redefinition of what is within the competence of the state and what is not, the public versus the private, and so on. Thus the state can only be understood in its survival and its limits on the basis of the general tactics of governmentality' (Foucault, 1978/2000, p. 221). Those tactics continually redefine and reform the purview of the state and so the state itself is a by-product of tactics of government and so also the state is, ultimately, less important than the tactics of government that comprise the state's appearance as a coherent entity (Mitchell, 1991).

Another way to put this is to say that changing what counts as political, and thereby what falls within the jurisdiction of state action, both sustains and transforms the state. Rather than an absolute limit and container of all possible politics, the state is constantly changing and is, like the prison, the product of political contestation over what is properly political, over what is changeable and what is not. Hence, reformers debate how far it is possible to transform the character of the criminal or the welfare recipient, or if it is possible to do so at all, and the answer determines in large part what will be undertaken.

The governmentalization or reform of the prison, public health, education, and welfare, for example, are not *strategies* on the part of the state to control the population and obtain the willing acquiescence of the people. Rather, they are the *tactics* – 'both internal and external to the state' – for changing what is or is not political, what is or is not within the purview of government (what I refer to below as 'neopolitics'). As such, they indicate just how variable the state is, how porous and indistinct its boundaries and limits are, and hence how indirect must be the path for transforming the state. Agitation for reform, then, is never *merely* reform or co-opted resistance to the state, as opposed to a more revolutionary or radical transformation of the state. Reform is the means by which the state is transformed *and* the source of its stability. Rather than an instrument for depoliticizing state power, reform (or in Foucault's terms, the 'governmentalization' of the state) is a means at our disposal to change what counts as politics and so to produce new conditions of political action, both internal and external to the state. By concentrating upon tactics of government (reform) rather than the state, as Foucault and many others have pointed out, the state is displaced as the limit-effect of political action.

His conception of governmentality carries the promise of freeing reform from its opposition to revolution for it is the relationship of political action to the state that anchors the opposition. Unleashing reform from the state can be accomplished not by reconceptualizing the state in order to revolutionize it, but by reconceptualizing politics as a mode of action that does not rely upon overcoming history and the limits placed upon action in the present, both highly improbable feats in any case. My argument here is that politics and political action are available to us in ways that have been under-appreciated and the primary form of political action I have in mind is adapting or changing what counts as politics, what I call 'neopolitics'. Reconceptualizing politics as a constantly shifting terrain, and I am able to offer only the barest inkling of such a monumental undertaking in this chapter, does not guarantee a more radical form of action, but it does contribute to understanding why even our best efforts to open the door of politics to radical movement and change are so disappointing. I pursue a 'continuist' thesis here that reform is perpetual, stabilizes the state, and yet holds radical promise.

Recall that the politically useful effect and productivity of disciplinary power, according to Foucault, was not to eliminate crime, but to distinguish and produce the criminal whom political power might exert itself upon, contain, and possibly reform. 'For the observation that prison fails to eliminate crime, one should perhaps substitute the hypothesis that prison has succeeded extremely well in producing delinquency, a specific type, a politically or economically less dangerous – and occasionally useable – form of illegality' (Foucault, 1979, p. 277) The reform of the prison gives birth to new kinds of subjects and to new tactics of government. The delinquent is a political subject, an effect of governmentality, who eventually gives way to the 'at risk teen,' the 'recidivist,' 'the sexual offender,' or the 'super predator'. The tactics of government may change from an ethnically-based criminality to an urban-spatial one, or from, say, a professionalized police to 'community policing'. The same may be said for charity, philanthropy, welfare, family, education, and health. Every reform, as certain subjects and tactics come into or fall out of play, also introduces a distinctive politics all its own, attended by new expertise, new tactics, and new subjects. The apparent stability of the liberal state is undergirded by a form of politics that is itself undergoing constant transformation, adaptation, and change.

As Paul Veyne expresses this, 'We need to turn away from standard 'politics' to notice an *exceptional* form, a political period piece whose surprising convolutions constitute the key to the enigma [of gladiatorship and its suppression]' (1997, p. 149). As a long series of exceptional forms, the political history of the liberal state is best expressed as 'neopolitics,' or, politics in a state of adaptation and change. 'Politics' has no more constancy or explanatory power than 'the state' and does nothing to explain how resistance is played out, overcome, or why things change. Veyne offers an example: it is common to assert that there have always been the governors and the governed, but 'let us consider the fact that practices for dealing with "the governed" may vary so widely over time that the so-called governed have little more in common than the name' (1997, p. 150). A 'criminal,' or any political subject, like the tactics used to govern that subject, is variable over time and the level of abstraction required to use 'the liberal subject' or 'the state' as explanatory of history is misleading in the extreme. Thus, 'reform' too should not be understood to be one and the same across time. For example, where once the only practicable reform was social reform (distinguished from economic reform or political reform or religious reform), so at another time, reform is distinguished in some other way. Enormous energy must be expended to provide genealogical accounts of reform and politics to fully bring their contingency and variability clearly into view. Still, no matter how profoundly contingent our concepts and our values may come to appear to us, and no matter how dramatic the shift in our perception of them, they remain sedimented in the discourses and practices that make up our way of life, in the political desires that move us, as well as in the political practices by which we are governed and by which we govern ourselves.

A 'continuist' rejects both the idea that history repeats itself and the idea that a complete break with the past is possible. While it is true that reform of the prison is perpetual, that should not be taken to mean that all reform is repetitive or serves the same purpose. When we hear an echo of the past in a charge that the prison is too harsh or too lenient (or that welfare produces dependency or punishes the poor, that schooling is too rote or too variable, that the family is too authoritarian or too disordered), we should beware that rhetorical echo does not deafen us to the sound of very real changes afoot (cf. Hirschman, 1991). By the same token, and perhaps this is more relevant today, as we shall see, when we hear the clarion call that a new day is upon us, we should beware that the political problems of the past are never behind us, but constitute the politics of the present.

## Neoliberal reform: instrumentalizing individual freedom, or the end of the willing self?

Between the spectacle of sovereign power and the modern timetable of disciplinary power, Foucault drew a stark contrast between historical modes of subjection and resistance. Yet once the prison was instituted, the cycle of reform continued to produce an acutely liberal mode of subjection, one that instrumentalized the individual will and relations of power. One could say that the political impetus behind Foucault's genealogies was to make those reforms less automatic, to challenge the self-evidence of the prison, and more generally to call into question the rationality of liberal and reformist government itself. That is why he refused to offer alternative schemes of government or proposals for what should be done. 'But my project is to bring about that they "no longer know what to do," so that the acts, gestures, discourses that up until then had seemed to go without saying become problematic, difficult, dangerous. This effect is intentional.' (Foucault, 2000, p. 235). Perhaps his intentions have been realized. Although I pursue a continuist thesis here, there is alarming evidence that presses against my thesis.

Consider the fact that today in the United States there are some two million prisoners, almost five million on parole, and undisclosed numbers held in undisclosed detention camps around the world. Overcrowded prisoners are abandoned to the order imposed by prison gangs or isolated day and night under video surveillance; prisons and prison services are privatized and contracted out. Prisons today appear to be warehouses for big business ('the prison industrial complex') more than political institutions for the discipline and reform of individuals. The self-evident character of the prison, 'which we find so difficult to abandon,' according to Foucault, is based on the deprivation of liberty and 'its role, supposed or demanded, as an apparatus for transforming individuals' (1979, p. 233). It may seem that today, however, the transformation of the individual is wholly supplanted by the deprivation of their liberty. It is tempting, and as we will see, quite commonplace to hear that the present forms of governmentality, and neoliberalism in particular, signal a dramatic break with the past and that the present marks the end of perpetual reform, if not political change altogether.

On the other hand, it is also possible to see neoliberalism from the vantage point of a 'continuist,' wherein Foucault's aims to stymie the perpetual reform of the liberal state are *not* realized. In this view, the contemporary privatization, deregulation, and decentralization of the prison (as well as welfare programs and education) does not signal the unleashing of a new or more ominous form of power, but re-establishes the conditions for the functioning of the prison by posing the harsher terms of imprisonment as the solution to the problems of punishment, to its excessive costs, recidivism, its previous leniency, and so on. Foucault suggested that liberal governmentality in general swings between the critical poles of governing too much or too little and the continuity of liberal government is guaranteed by its reform or rebalance between these two poles. Neoliberal reforms are interpreted in the literature on governmentality, inspired by Foucault, as re-tipping the balance away from a welfare state that governs too much. On this view, the neoliberal ' "retreat from the state" is also itself a positive technique of government; we are perhaps witnessing, a "degovernmentalization of the State" but surely not "de-governmentalization" *per se'* (Barry, Osborne and Rose, 1996, p. 11). Hence the retreat from the state necessitates a reformed art of government (privatization and decentralization) and subjectivization (the enterprising self), but not an entirely new kind of subject or power. Neoliberal reforms, then, might be understood to confirm Foucault's thesis that reform remains a principle of liberal rule.

As others have amply demonstrated, neoliberalism does not simply free the will of individual economic actors from government restraint or free market forces from state intervention. Rather, neoliberalism seeks to shape and instrumentalize individual freedom. As Nikolas Rose writes, 'Within this new regime of the actively responsible self, individuals are to fulfill their national obligations not through their relations of dependency and obligation to one another, but through seeking to fulfill themselves within a variety of micro-moral domains ...' (1996c, p. 57). Micro-credit schemes are a prime example of the governmentalization of economic action outside the domain of the state and directed largely by NGOs, international aid agencies, and non-profit enterprise (Goldstein, 2001). The individual impoverished subject is transformed and reshaped into an enterprising self whose well-being is no longer the object of state obligation. The privatization of government, then, does not entail less government, but government in a new domain with a new kind of subject. Rather than simply abandoning the willing self to the market, neoliberalism aims to reshape the willing self into a market actor. The continuist thesis pursued here is still not without its challenges, however.

There is evidence that at least in the United States, we *have* abandoned the disciplinary project of transforming the individual even if the prison still stands. Life sentences, three-strikes laws, administrative detention, the decline of the welfare state, and privatization of education, all appear, and their proponents even proclaim, to abandon the possibility of individual reform, particularly the reform of people of color. Is it time to rethink if not reject a 'continuist' view of the liberal state? To be sure, neoliberal reforms are so punishing, and intentionally so, it may well be the case that the will, autonomy, and freedom of the governed are no longer useful as the instruments and effects of governmentality.

Do neoliberal policies signal that the era of perpetual reform of the liberal tactics of government is over? Is there reason to suspect that the liberal state is no longer sustained by tactics of government and their perpetual reform or, perhaps, that the state is no longer liberal? Have modern forms of power including discipline been supplanted by new and more ominous forms of power? One obvious reason to consider these possibilities is that so many commentators declare that we are, in fact, at the end of modernity and the era of perpetual reform. I consider three such claims: that the present marks the end of history, of disciplinary society, and the end of liberal democracy.

## The end of politics?

Neoliberal policies of privatization, deregulation, and state decentralization have spawned a chorus of contention that, whether celebrating or grieving the difference of the present from the past, proclaims 'the end of history,' 'the end of left politics,' and 'the end of liberal democracy'. The chorus heralds the birth of a new era of postmodernity that is variously labeled post-industrial, post-scarcity, post-welfare state, an era that is post-democratic, post-social, post-disciplinary, post-colonial, or post-9/11. The chorus situates the present in terms of the past, as coming after ('post'). The present is pure dissimilitude, known, named, and inaugurated by its dissimilarity to the past, by what is threatened, displaced, lost, or overcome.

Gilles Deleuze argues that disciplinary society has given way to a 'society of control' or a 'post-disciplinary society,' not as a consequence of reform, but by the breakdown of disciplinary institutions of confinement. In a disciplinary society, individuals 'are always going from one closed site to another, each with its own laws: first of all the family, then school ... then the barracks ... then the factory, hospital from time to time, maybe prison, the model site of confinement' (Deleuze, 1995, p. 177). In 'societies of control,' by contrast, a new system of domination is continuous and its model is individual debt. One could say the life sentence to prison and three-strikes laws are also exemplary of the new system of continuous control. Disciplinary tactics and institutions are expensive both in terms of resistance and in terms of the costs associated with containing populations within the confines of family, factory, school, prison, and hospital. Discipline today, by contrast, is privatized; individuals are responsible for their own discipline and we pay dearly for private schools, personal trainers, life coaches, career training, and rehab. Prisoners and their allies must organize and fight for the 'privileges' of discipline from physical work outs and educational opportunities to drug counseling. What is so important about this for Deleuze is not that what was once public is now private, but that the form power takes is unique: control and continuity instead of discipline and enclosure.

Does the privatization of discipline reflect the further governmentalization of the state, or does neoliberalism work by disbanding the state and its institutions and abandoning the individual to the private sector? It may appear that the individual will is controlled rather than instrumentalized in the neoliberal state and that the post-disciplinary society commands ('abstain!', 'choose!', 'Just say No!', 'Just do it!') and coerces rather than disciplines. Indeed, in the United States, we have bridefare instead of daycare, workfare instead of jobs, abstinence-only instead of safe-sex education, mandatory school testing instead of truancy officers, family values and child support enforcement instead of nuclear families, electronic bracelets instead of parole officers, personal trainers instead of sports, home work instead of factories, home care instead of hospital care, pharmaceuticals instead of psychotherapy, video surveillance instead of police, volunteer instead of conscription army, Defense of Marriage Act instead of family policy, and community- and faith-based social services rather than state bureaucracy. In every case, through privatization and deregulation, the power of the state is 'devolved' or 'decentralized' and it disperses power in a way that is not spatialized or institutionalized in state institutions.

However, it is also possible to construe the same trends as the governmentalization of private life, voluntary action, and the community (Rose, 1996b; Hyatt, 2001). That signals an extension of neoliberal governmentality into new domains and not its end point. Individual 'choice,' personal responsibility, volunteerism, and the zones of intimacy fall within the purview of government and self-government. For example, people on public assistance of any kind, even though they no longer have a right or entitlement to state welfare since 'the end of welfare as we know it,' are still schooled in the morality of family values, sexual abstinence, marriage, enterprise, work, character, and so on (see Mink, 1998; Schram, 2000). However, it may not be state institutions or agencies working to transform the individual will. It is more likely to be a private, non-profit or for-profit agency, a community-based organization (CBO) or a faith-based organization (FBO), with or without state funding, working to transform the individual will (see Minow, 2003). Rather than privatizing the state, we are witnessing the governmentalization of individual will power in private non-profit enterprises. The ground of politics is shifting from society to the willing self. Today we take up our citizenship not in relation to the state, but in private or in civil society.

As totalizing and violent as neoliberal tactics may be, it is important to recognize that the very ground of politics is moved and with it the possibility to ground new practices of freedom at the level of the will and the community (see Gibson-Graham, 2006). To bring that ground into view, it is necessary to contend with other voices in the chorus of contention who insist that there are no ideological or normative alternatives to neoliberalism as evidence that we are at the end. Francis Fukuyama and Wendy Brown, for very different reasons and from different points in the political spectrum, declare the present to be the end of history and liberal democracy, respectively. They mean much the same thing. In a famously celebratory mode, Fukuyama (1992) prophesied the triumph of 'the Western idea' (liberal capitalism). With the dismantling of the 'socialist alternative' to capitalism, we are delivered to 'the end of history'. All political possibilities are foreclosed by the triumph of liberalism. 'The end of history will be a very sad time. The struggles for recognition, the willingness to risk one's life for a purely abstract goal, the worldwide ideological struggle that called forth daring, courage, imagination, and idealism, will be replaced by economic calculation, the endless solving of technical problems, environmental concerns, and the satisfaction of sophisticated consumer demands' (Fukuyama, 1992, p. 18). Fukuyama proclaims that the neoliberal variant of liberalism is the final moment in the progress of history. Without an ideological alternative to liberalism, according to Fukuyama, we will no longer *make* history; we are fecklessly redeemed *from* history and from politics.

Like Fukuyama, for Brown, we are at a point in history when new political possibilities and alternatives to the present are rendered impossible by the historical conditions of the present. In Brown's essay (2003), 'Neo-liberalism and the End of Liberal Democracy', however, it is not the inevitable march of history and the progress of liberal ideology delivering us to our fate, but the constitutive economic rationality of neoliberalism that destroys the possibility of action. 'While this entails submitting every action and policy to considerations of profitability, equally important is the production of all human and institutional action as rational entrepreneurial action, conducted to a calculus of utility, benefit, or satisfaction against a micro-economic grid of scarcity, supply and demand, and moral value-neutrality' (Brown, 2003, p. 9). Neoliberalism *produces* rather than presumes the end of history. As a consequence, democracy is 'displaced' and democratic citizens are reduced to *homo economicus* by neoliberalism's relentlessly depoliticizing rationality. 'Liberal democracy,' she proclaims, 'cannot be submitted to neo-liberal political governmentality and survive' (Brown, 2003, p. 23).

In other words, neoliberal governmentality is the end of liberal democracy. Rather than continuous with liberalism, neoliberalism 'eviscerated' the moral and political underpinnings of liberal democracy, according to Brown. On the one hand, neoliberalism dries up or displaces the democratic conditions for developing 'a counter-rationality' to neo-liberal reforms, and on the other hand, such a counter-rationality is the only thing that will save us from the conditions of the present. Rather than prophetic and complaisant like Fukuyama, Brown's jeremiad is a call to action in the face of the disappearance of the conditions of resistance. 'What remains for the Left, then, is to challenge emerging neo-liberal governmentality in EuroAtlantic states with an alternative vision of the good, one that rejects *homo economicus* as the norm of the human *and* rejects this norm's correlative formations of economy, society, state and (non)morality' (Brown, 2003, p. 42). Brown warns us away from the melancholy that Fukuyama embraces and would have us embrace instead a counter-rationality, an alternative vision of democracy: 'In its barest form, this would be a vision in which justice would not center on maximizing individual wealth or rights but on developing and enhancing the capacity of citizens to share power and hence, collaboratively govern themselves' (Brown 2003, p. 42). Such an alternative vision does not yet exist and thus, the perpetual reform of liberal institutions and the ground of critique are permanently displaced by neoliberal market rationality.

Brown and Fukuyama share the view that we are reduced to economic activity after the triumph of neoliberalism. Politics itself is done. However, there is a great deal of historical evidence that economic rationalities and market calculations were applied to political problems long before the advent of neoliberalism. Brown ignores the role of economic thought in the long history of liberal democratic reform in factories, welfare, prisons, a long history of economic rationality in evangelical schemes and benevolent work. I will cite just two examples. Lori Ginzberg notes that at the time of the Civil War, 'wartime benevolent work celebrated explicitly business, not evangelical, principles . . . [which] signaled not a millennial vision of society but a glorification of the new virtues of efficiency and order' (1990, p. 133). As Nikolas Rose points out, '[F. W.]Taylor too sought to forge an image of work consistent with the values of democracy.

The real basis of industrial democracy, he argued, is the "grassroots democracy" established by effective communication, appropriate managerial attitudes, and so forth' (1996, p. 147). It is not as if all of a sudden, with the advent of neoliberalism, we are suddenly subjected in an economical way that completely destroys our political freedom, even if it is a way that instrumentalizes our freedom differently. On my view, to govern economically is and was linked to rather than displaced by the will to govern democratically and liberally.

Our liberty continues to be both an effect and an instrument of neoliberalism with the discipline of the market aiming to supplant the discipline of state institutions: to free the market from the regulation of the state, to govern better, the state must govern less. Neoliberalism both presumes and produces the individual will freed from governmental restraints. The vision of contemporary neoliberalism is that markets will lead to peace once they are relieved of state regulation. The market has a pacifying, moralizing, democratizing, and homogenizing effect. Neoliberals reject the Keynesian vision of the state as protector, regulator, and cleaning crew of the economy. The market stands in the stead of the Keynesian state: tax cuts, deregulation, privatization, enterprise. It is self-regulating (all scandals aside) and itself a regulatory principle for motivating voluntary action. Neoliberal policies rely upon the autonomy and economic rationality of the individual will to replace governmental functions. Or so both the advocates and the critics of neoliberalism say.

Like Brown, I prefer democracy to any variant of liberalism and seek for sources of resistance to the instrumentalization of the willing self. Yet an alternative democratic vision is blind without its own rationality of government for, as Brown puts it in the quote above, 'enhancing the capacity of citizens to share power and hence, collaboratively govern themselves'. For such a vision to be realized, politics must move, again, from the willing self, the market, and civil society under neoliberalism, to some new terrain. I remain optimistic that some reform scheme may well get us there, yet am doubtful that an alternative vision of the good is adequate. Fukuyama promises and Brown fears that citizens and nations alike will be consumed in and freed by their economic activity and stop making so much trouble. However, each of their arguments hinges upon the absence of alternatives to neoliberalism, which there is every reason to doubt.

## Neoconservative reform: giving up on the willing self?

Politics persists and at least one new alternative to neoliberalism is in ascendance. Neoconservativism in the United States combines communitarianism and neoliberal market rationality. Neoconservativism diverges sharply from neoliberalism and contests that the will and its sources are so corrupted by liberalism and multiculturalism that to lift the restrictions of the state upon the will is to unleash a dangerous and unpredictable force. To be sure, neoconservatives do not reject the vision of neoliberalism. Irving Kristol, the 'grandfather of neoconservatism,' wrote: 'It is a basic assumption of neoconservatism that, as a consequence of the spread of affluence among all classes, a property-owning and tax-paying population will, in time, become less vulnerable to egalitarian illusions and demagogic appeals and more sensible about the fundamentals of economic reckoning' (Kristol, 2003). The problem for neoconservatives is that market rationality cannot produce the moral ground on which it stands or falls, by which it produces general affluence or inequality and poverty. The liberal welfare state had the perverse effect of being literally socially constructive: it *caused* poverty by allowing women to become single parents incapable of acting economically and rationally. The poor developed a 'culture of poverty and dependency' in response to a welfare state that was 'too permissive'. The welfare state had to be undone, not only to free market forces, but also because it was producing irrational effects in the form of pluralizing 'traditional values,' promoting single-parenthood, de-incentivizing the 'traditional family,' with particular emphasis on minority statistics. According to neoconservatives, the welfare state took over too many functions better left to voluntary sectors, and that cannot, as neoliberals believe, simply be provided by the market. Neoconservatives warn that there is no market incentive to sustain monogamous marriage or 'traditional values'. Civil society, neoconservatives contend, is in a shambles because of the 'sexual revolution,' feminism, gay rights, multiculturalism, civil rights, and so on. Civil society and the market are powerless to impose their own rationalities upon the corrupted will of the people. According to the neoconservatives, only a strong and decisively illiberal state can work that magic.

Gertrude Himmelfarb, a neoconservative, puts it this way: 'Those who want to resist the dominant culture [by which she means the culture of permissiveness, multiculturalism, postmodernism, and feminism] cannot merely opt out of it; it impinges too powerfully upon their own lives and families. They may be obliged, however reluctantly, to invoke the power of law and the state, if only to protect those private institutions and associations that are the best repositories of traditional values' (1994, p. 248). In other words, we need to adopt illiberal reforms to 'remoralize' the will of the people and their dominant culture. Willing selves are not born, they are made. Neoliberalism goes too far in down-sizing the state.

Liberty will be assured only after the state lays the ground for economically rational action by 're-moralizing' the people. Neoconservatives understand their alliance with the religious right as reconciling economic conservatism with social and cultural conservativism. Thus, the terrain of state action has to shift from the 'welfare of the people' guaranteed by the economic rationality of neoliberalism, to waging war in the domain of culture by an authoritarian state. Rather than limit state power, neoconservatives seek to combine the deregulation of the market with authoritarian state action.

They propose to use state and juridical power to enforce the norms of sexual abstinence until marriage, the norm of marriage itself, work, voluntarism, and political participation, because the will of a corrupt people cannot be used to its own advantage. That does not yet mean fundamentally shutting down the regulatory arms of the state as so many traditional conservatives wanted. The 'new paternalism' heralded by Lawrence Mead and the 'compassionate conservatism' of President Bush combine to prevent the complete dismantling of state welfare provision and downsizing of government.

Bush, for example, contrasted the failed compassion of state bureaucracies to the authentic compassion of faith-based organizations, charities, and community groups, but he did not deliver an anti-government conservative agenda (Apostolidis, 2001). By governmentalizing compassion, love, and intimacy in faith-based organizations, for example, the territory of government is moved rather than simply contracted.

Calling for a stern 'new paternalism', Mead advocated workfare programs as a means of governmentalizing individual responsibility: 'As every political theorist – and policeman – knows, government can rarely control people if it merely blames or coerces them. Rather, citizens must accept its demands as legitimate, transmuting mere power into authority. Only then can compliance be widespread. Far from blaming people if they deviate, government must persuade them to blame themselves' (Mead, 2001, p. 10). Obligation cannot be successfully imposed upon the poor by the state. Rather, the poor must come to take responsibility upon themselves to work and that entails 'public norms and enforcement that are collective in character' (Mead, 2001, p. 10). It is, in other words, a political solution, as Mead insists, and not a technical or policy solution that is required to make the poor responsible for themselves. Rather than dismantling the welfare state, its entitlements and rights are dismantled, and new obligations and personal responsibility are imposed upon the poor. It is disciplinary in a new domain of government, but it is still disciplinary.

Internationally, neoconservativism talks tough and aims to impose democracy, freedom, and free markets abroad. Neoconservatives are conservative interventionists, an oxymoron by traditional conservative standards. Neocons see their national interest in the (now globally familiar) terms laid out by Irving Kristol. 'Finally, for a great power, the "national interest" is not a geographical term, except for fairly prosaic matters like trade and environmental regulation . . . And large nations, whose identity is ideological, like the Soviet Union of yesteryear and the United States of today, inevitably have ideological interests in addition to more material concerns.' He adds, 'No complicated geopolitical calculations of national interest are necessary.' Our prerogative, in other words, is to distinguish friend from foe without any justification other than knowing our ideological bedfellows.

In Foucauldian terms, neoconservatives contend that the liberal subject is an effect, but no longer viable as an instrument of government in the Arab world, no less than in the demoralized West. Hence the liberty of the people must be constrained and re-moralized rather than maximized. Neoconservatives favor a strong state and believe that rather than homogenizing the population, free markets alone contribute to the moral decline and diversification of the population. New markets produce new and distinctive populations, desires, and rationalities. Liberalism and disciplinary power fail, in their view, as an apparatus for transforming individuals because they are 'de-moralized'.

Neoconservativism and neoliberalism are distinct ideologically, but they are practically wedded in the policy and legislation of the United States. For example, abstinence-only policy at home and abroad is pursued not by reforming comprehensive sex education, but by abolishing it and substituting abstinence-only commands (appearing perhaps to be the tools of what Deleuze calls a control society). As part of the Personal Responsibility and Work Opportunity Reconciliation Act (1996) that famously 'ended welfare as we know it,' in President Clinton's words, and ended welfare entitlements by 'devolving' federal responsibility for welfare to the states, abstinence-only introduced an eight-point 'educational or motivational program'. To summarize those eight points, the programs must teach that avoidance of disease and 'out-of-wedlock pregnancy' can only be assured by sexual abstinence. Provision F teaches that sexual activity outside of the context of marriage is likely to have harmful psychological and physical effects. It is referred to by critics as the 'fear curriculum'. Is this government by control, mandate and fear rather than discipline? I think not. It is the governmentalization of individual sexual practices. One becomes a responsible and upstanding citizen, one 'entitled' to welfare benefits, childcare, and work training only by virtue of one's own self-control or signing a 'chastity pledge'.

The bill made provisions for non-governmental organizations (NGOs) like FBOs (faith-based organizations) and CBOs (community-based organizations) so that even states would be by-passed in the provision of welfare and social programs, with churches and non-profits providing programs for teaching sexual abstinence, 'the promotion of marriage,' and implementing workfare programs. Government is decentralized, so to speak, and civil associations and church congregations are governmentalized. In this respect, neoconservative programs are consistent with the privatizing and market trends of neoliberalism. These programs have the appearance of being authoritarian and command-based, not apparatuses or technologies for transforming individuals, but, in fact, they extend tactics of government into new domains of civil society to moralize and shape the will of the individual. Of course, the story of perpetual reform I am telling here is also a story of the perpetual failure of tactics like neoconservative and neoliberal variants of governmentality. These programs too will fail to re-moralize and instrumentalize the willing self along the traditionalist lines neoconservatives envision.

Rather than detailed analysis of these programs, I present neoconservative and communitarian programs here as alternatives to neoliberal governmentality and as such, they portend that the era of perpetual reform and liberal governmentality is still far from its end. Still more, it is not a return to the past, as (tough talking) neoconservative and (civil) communitarian proponents like to claim. In Foucault's terms, we might ask, is the 'promotion of marriage' a return from the deployment of sexuality to the deployment of alliance? Are faith-based initiatives and workfare a return to an earlier pastoral power? However, these questions presuppose some fixity to the confusion and complexity of competing schemes of government that characterize the present.

The hand-wringing over the 'lack of alternatives' to (neo)liberalism betrays a desire for finality to silence the cacophony of politics in the present. The chorus calling the present into question bears some resemblance to what Michel Foucault described as the ethos of modernity, the critical ontology of ourselves, or the engagement in a kind of permanent critique of the present and of ourselves in relation to the present. Yet the chorus sings of its own difference from the past, not its identity in the present. Foucault suggested that the ethos of modernity 'is the will to "heroize" the present' (1984, p. 40). By that, he meant the will to realize the present by affirming its limits. The chorus of contention refuses or fails to heroize the present because its limits are set by the past. The ethos of postmodernity, you could say, is the will to 'heroize' the past. In the chorus of contention that we are on the brink, the critical attitude toward the present is displaced by the question of our dissimilarity to modernity itself. Does the emergence of this new 'ethos of the end' signal that the critical attitude that guaranteed liberal governmentality is past? In this context, we must ask again if the perpetual reform Foucault described continues to be the guarantee of liberal forms of rule? Put somewhat differently, does the dismantling of welfare states and the privatization of governmental functions signal the end of disciplinary power and the perpetual reform that guaranteed its exercise?

## Neopolitics of the present

An alternative to situating the present at the end, and perhaps a better way to approach the question of reform, is to embrace neopolitics and a 'continuist' vantage point on the history of the present. By that I certainly do *not* mean to embrace neoliberalism and neoconservativism. Rather, I mean to embrace the present as the time in which politics is the practice of contesting the rationalities, ideologies, the organization of power, and the freedoms of subjects, in the past. One way to characterize the present, then, is that it is accompanied by the chorus of contention over our difference from the past. Our politics have become, in a sense, genealogical, and the contingency of politics itself calls out for remedy. To situate the present at the end and to declare the present 'post,' is little more than the futile effort to remedy the condition of neopolitics, to fix the present into a relation with the past. This has the unfortunate effect of closing down upon new political possibilities and abandoning the willing self to neoliberal instrumentalization or the moral economy of neoconservatives.

Rather than the end of politics, a continuist view renders the exceptional form of politics in the present into a moment of possible action. The neopolitics of reform gives name to our acute consciousness of the contingency of the political and, hopefully, provides an indication that it is possible to open the present to new political possibility. The 'continuist' vantage point is not for or against the past, but allows us to adapt to the knowledge we have of how uncertain the past was and remains. The dividing line between revolution and reform is a place to begin. Even if reformers are selling themselves out to the state, they are nevertheless revolutionary in the only way possible, that is, to change or adapt what counts as politics. They are, if you will, neopoliticians.

It is also possible to take advantage of the neopolitics of reform and to deploy tactics of government toward more participatory and democratic ends. An opening for a more democratic movement of reform is coming into view, one that might take advantage of the ways the contingency of politics is revealed in the present. Rather than an alternative vision of the good to combat the authoritarian and traditionalist neoconservative or the triumphalist neoliberal, those of a democratic faith require tactics of government to shape citizens capable of self-government. The willing self, even a democratically constituted one, is made and not born. A democratic governmentality would deploy the tactics of government to constitute democratic citizens capable of contesting and administering their own constitution (cf. Olson, 2006). And democratic reform will fail, of course, because the political ground will shift and the cry for new reforms, more democratic measures, and a renewed citizenship will arise. In the continuist view, failure is no reason to stop trying. Failure is the very condition of success for the ongoing activity of politics, for grounding the practice of freedom in politics.

Foucault's conception of governmentality wedded relationships of power (parent/child, manager/worker, warden/prisoner, and so on) to rationalities of government. 'Neopolitics' distinguishes the political dynamics within governmentality, the dynamics that simultaneously contain political conflict within the state and transform what can be counted as political and, hence, fall within the purview of government. However, neopolitics does not guarantee a more radical form of action. The neopoliticians of the present (neoliberal, neoconservative, and post-modern) are (for better and for worse) in the business of transforming politics, moving out from the state and into the community, the market, the personal and intimate, and civil society. With the privatization of the state and 'the personal is political,' from above and from below, the self-evidence of the state's monopoly on politics is now in question. It seems quite possible in the moment, then, to begin to unleash politics and political will from the state and thus to unleash reform from revolution. Reform is a practice of freedom, and a special one at that, for it is a practice of changing the very ground of freedom and determining what is or is not actionable, what is or is not within our grasp to change.

Neopolitics is not an alternative to the present or a manner of closing down upon the past, but a manner of situating the present in relation to the past that opens up new political possibilities. The present does not express the definite shape of the future anymore than that of the past. Rather, neopolitics expresses the limits and uncertainties of the present, its openness to new configurations of power and subjectivity, territory and governance. Once its contingency is revealed, politics will never be the same; but then again, it never was.

# References

Apostolidis, P., 'Homosexuality and "Compassionate" Conservatism in the Discourse of the Post-Reagan Right', *Constellations*, VIII, 1 (2001): 78–105.

Barry, A., T. Osborne and N. Rose (eds), *Foucault and Political Reason: Liberalism, Neo-Liberalism, and Rationalities of Government* (London: University College London Press, 1996).

Brown, W., 'Neo-Liberalism and the End of Liberal Democracy', *Theory & Event*, VII, 1 (2003). Available at: http://muse.jhu.edu/journals/theory_and_event/v007/7.1brown.html

Deleuze, G., 'Postscript on Control Societies', in G. Deleuze, *Negotiations: 1972–1990* (New York: Columbia University Press, 1995), pp. 177–82.

Foucault, M., *Discipline and Punish: The Birth of the Prison* (New York: Vintage Books, 1979).

Foucault, M., 'Questions of Method', in J. D. Faubion (ed.), *Power: Essential Works of Michel Foucault, 1954–1984* (New York: The New Press, 1980/2000), pp. 223–38.

Foucault, M., 'What Is Enlightenment?', in P. Rabinow (ed.), *The Foucault Reader* (New York: Pantheon Books, 1984), pp. 32–50.

Foucault, M., 'The Ethic of Care for the Self as a Practice of Freedom', in J. Bernauer and D. Rasmussen (eds), *The Final Foucault* (Cambridge, MA, and London: The MIT Press 1987/1991), pp. 3–12.

Foucault, M., 'Governmentality', in J. D. Faubion (ed.), *Michel Foucault: Power* (New York: The New Press, 1978/2000), pp. 201–22.

Fukuyama, F., *The End of History and the Last Man* (New York: The Free Press, 1992).

Gibson-Graham, J. K., *A Postcapitalist Politics* (Minneapolis: University of Minnesota Press, 2006).

Ginzberg, L., *Women and the Work of Benevolence: Morality, Politics, and Class in the 19th-Century United States* (New Haven and London: Yale University Press, 1990).

Goldstein, D. M., 'Microenterprise Training Programs, Neoliberal Common Sense, and the Discourses of Self-Esteem', in J. Goode and J. Maskovsky (eds), *The New Poverty Studies: The Ethnography of Power, Politics, and Impoverished People in the United States* (New York and London: New York University Press, 2001), pp. 236– 72.

Himmelfarb, G., *The De-Moralization of Society: From Victorian Virtues to Modern Values* (New York: Vintage Books, 1994).

Hirschman, A. O., *The Rhetoric of Reaction: Perversity, Futility, Jeopardy* (Cambridge, MA, and London: The Belknap Press of Harvard University Press, 1991).

Hyatt, S. B., 'From Citizen to Volunteer: Neoliberal Governance and the Erasure of Poverty', in J. Goode and J. Maskovsky (eds), *The New Poverty Studies: The Ethnography of Power, Politics, and Impoverished People in the United States* (New York and London: New York University Press, 2001), pp. 201–35.

Kristol, I., 'The Neoconservative Persuasion', *The Weekly Standard*, VIII, 47 (25 August 2003).

Mead, L., *Beyond Entitlement: The Social Obligations of Citizenship* (New York: Free Press).

Mink, G., *After Welfare* (Ithaca, NY: Cornell University Press, 1998).

Minow, M., *Partners, Not Rivals: Privatization and the Public Good* (New York: Beacon Press, 2003).

Mitchell, T., 'The Limits of the State: Beyond Statist Approaches and Their Critics', *American Political Science Review*, LXXXV, 1 (March 1991): 77–96.

Olson, J., *Reflexive Democracy: Political Equality and the Welfare State* (Cambridge, MA, and London: The MIT Press, 2006).

Rose, N., *Inventing Ourselves: Psychology, Power, and Personhood* (Cambridge and New York: Cambridge University Press, 1996).

Rose, N., 'The Death of the Social? Re-figuring the Territory of Government', *Economy and Society*, XXV, 3 (1996b): 327–56.

Rose, N., 'Governing "Advanced" Liberal Democracies', in A. Barry, T. Osborne and N. Rose (eds), *Foucault and Political Reason: Liberalism, Neo-Liberalism, and Rationalities of Government* (London: University College London Press, 1996c), pp. 37–64.

Schram, S., *After Welfare: The Culture of Postindustrial Social Policy* (New York: New York University Press, 2000).

Veyne, P., 'Foucault Revolutionizes History', in A. I. Davidson (ed.), *Foucault and His Interlocutors* (Chicago: The University of Chicago Press, 1971/1997), pp. 153–73.

# 7

# 'Craving' Research: Smart Drugs and the Elusiveness of Desire*

*Mariana Valverde*

## Smart drugs and targeted governance

'Smart bombs' were introduced with much fanfare by the US military during the first Gulf War to allay fears about the political consequences of repeating Vietnam-style 'carpet bombing'. The bombs dropped by the US Air Force, CNN told the world, were so smart that they could find and destroy military installations without causing massive civilian casualties. Like smart bombs, 'smart' drugs are supposed to act selectively on particular targets; they are part of a new era of medical treatment, an era characterized by less invasive and less expensive surgery and other 'targeted' strategies.

The appeal of smart weapons, military or medical, is that if targets can indeed be isolated and acted upon, using intelligence of various kinds, then authorities will be able to unleash powerful new devices upon political or bodily 'hot spots' – while minimizing the collateral damage and attendant bad publicity that have marred the forward march of technical progress for both military and pharmaceutical interests.

In the sphere of the social, targeted governance has also become extremely popular. Targeted policing, for example, is all the rage in North America. The targeting can be geographically specific – involving focusing on 'hotspots', that is, places known to have been the site of problems in recent times, to the detriment of regular foot patrols spread evenly throughout the city. Targeted policing can also be risk-factor specific, such as the profiling carried out by immigration officers. Either way, as I have shown elsewhere, there seems to be a sense that the city or the nation as a whole can't or shouldn't be policed: instead, there should be a rational selection of high-risk spaces, high-risk people, or risk factors (Valverde and Mopas, 2003).

In social services too, 'social planning,' the utopia of post-World War II social democracy, now sounds anachronistic and even embarrassing.

Instead of state-wide social planning (involving the development of universal programs) or even city-wide social planning, we have 'targeted social programs'. The idea is that social researchers will do their work and tell policy makers which are the 'high risk neighborhoods' or 'high risk populations,' so that resources can be allocated to those places or groups, rather than being distributed either universally or randomly.

One could see this, from a Foucauldian perspective, as the death of the dream of the panopticon. The panopticon ideal has received some bad press because of the noted practical – and legal – difficulties that have marred the efforts made by authorities, since the days of Jeremy Bentham, to exercise surveillance evenly over a whole territory and population. But I argue that 'targeted governance' is not always a modest proposal motivated by the recognition of the panopticon's failures: targeted governance is often touted not only as more practical or more respectful of privacy than universal governance, but as actually more desirable, normatively. In the health field, drugs that act systemically tend to be seen as inferior to drugs that act only on designated processes or on specific receptor sites in the brain. Governing the whole person or even the whole body is regarded not only as utopian but as frankly dangerous.

One of the reasons why targeted governance is taken to be inherently sound is that the whole notion of targeting requires specific information. Or, to put it differently, targeted governance is the form of power associated with such knowledge practices as 'evidence-based' medicine and 'intelligence-led' security.[1] The very language of smartness suggests, not unhelpfully for the authorities doing the targeting, that it is downright stupid to even attempt to observe the whole population and/or to provide services to the whole population. When we hear cabinet ministers, even in social democratic governments, tell the public that government must provide targeted programs for specific populations with specific needs, it is clear that the old welfare-state dream of universal and equal governance has died. So, too, when we hear medical authorities talk about abandoning whole-person projects (e.g., psychoanalysis) in favor of 'targeted' interventions, we can suspect that the key product of universalist humanist governance (Man himself) is perhaps also fading – as suggested in Foucault's evocative image of the 'face drawn on the sand at the edge of the sea' (Foucault, 1973, p. 387).

## Disorderly consumption and the problem of desire

I will explore the logic of targeted governance not in abstract theoretical terms but, as befits the topic, in relation to a particular target, specifically, an object that has long been a site for all sorts of discourses and governing techniques aimed at human willpower: alcoholism (Valverde, 1998). I will describe how the longstanding quest for a pharmaceutical magic bullet for alcoholism resulted in a cautious, but palpable euphoria about some drugs (developed for opiate addiction treatment, and thus known as 'opiate receptor antagonists') that are thought to have a largely unexplained but documented targeted effect upon excessive or uncontrolled drinking.

There are several opioid receptor antagonists that have demonstrated some success in relation to alcoholism in trials (e.g., nalfemene; see Sinclair, 2001). The most recent one of these is acamprosate, sold under the brand name Campral, which is too new to discuss here. For our purposes, which are theoretical rather than medical, we will focus on the opiate receptor antagonist with the highest profile in alcohol abuse treatment circles: naltrexone, which began to be marketed in pill form as ReVia in the mid-1990s and some years later in injectable form under the brand name Vivitrol.

The product announcement by DuPont for ReVia describes how this drug works:

> How does ReVia work? It is believed that alcohol causes the release of endogenous opioids. The binding of these opioids to the [opiate] receptors in the brain may be responsible for the positive reinforcing effects of alcohol. ReVia, an opioid receptor antagonist, competitively binds to these receptors, blocking the endogenous opioids at these sites.
>
> (cited in Sinclair, 2001, p. 2)

This product announcement carefully avoids using the stigmatizing psychiatric term 'dependence'. It thus leaves open the old question of whether drinking ought to be classified as an addiction/dependence or simply as a pleasure that has positive reinforcing effects and that has to be managed, like eating fast foods or driving fast cars. The agnosticism of DuPont's announcement thereby also circumvents the question that has plagued governments and public health authorities around the world and at the UN level, namely, the extent to which alcohol ought to be classified and regulated as a drug.[2]

Some medical studies do claim that naltrexone acts upon dependence itself (the term that has replaced 'addiction' in medical discourse); but, in keeping with the epistemologically modest logic of targeted governance, treatment practitioners and scientists using different theories can agree that naltrexone works to help people drink less even if they disagree on the larger issues of ontology and causation. For some, naltrexone and its relatives do act on addiction itself; for others, the pill merely suppresses what psychologists call 'craving' and thus indirectly helps people who want to decrease their drinking to do so.

Naltrexone was the first drug approved for alcoholism treatments by the US government since the approval 50 years ago of the now discredited, harshly disciplinary drug disulfiram (better known under its brand name of Antabuse). After taking an Antabuse pill, the alcoholic in treatment would become violently ill if he/she fell into the temptation of drinking. This kind of punishment might have succeeded in the very short run, but as a long-term treatment program it was clearly never going to work. Outside of the hospital or the 'drunk tank,' few people would voluntarily take their daily Antabuse pill.

Naltrexone started out in life about 25 years ago as a drug prescribed to help heroin addicts who wanted to quit. It is still used for that purpose, but its current main claim to fame is that it may turn out to be the Holy Grail of alcohol studies: an anti-alcoholism 'targeted medication' (Heinala et al, 2001; Sinclair, 2001). Rather than harshly punishing the drinker's whole body, as Antabuse does (a 'treatment' that now seems Dickensian), it targets the disorderly desire or disorderly willpower that is associated not with everyday social drinking but specifically with alcoholism. Antabuse is not smart, not only because it harshly punishes the body but also because it cannot distinguish between one civilized glass of brandy and a bender. Naltrexone, by contrast, targets the social and psychic problem associated with excess, immoderate, uncivilized drinking, otherwise known as alcoholism. The American National Institute for Alcohol Abuse and Alcoholism, which provides 90 percent of alcoholism treatment research funding in North America, announced the approval of naltrexone in the following words:

> Unlike disulfiram, naltrexone and other potential agents now under NIAAA investigation *directly target hallmark features of alcoholism*: abnormal alcohol-seeking behavior, impaired control over alcohol intake, and physiological dependence manifest in craving when alcohol is removed.
>
> (NIAAA, 1995; my emphasis)

This is reiterated in one of the most successful US private treatment programs using naltrexone, 'Assisted Recovery':

> While the precise mechanism of action for naltrexone is unknown, medical researchers believe that naltrexone blocks the ability of alcohol

to stimulate the release of endorphins ... thus effectively suppressing the craving or thought process to drink.

(ww.assistedrecovery.com, accessed 7 February 2007)

Naltrexone does not work by diminishing the effects of physical withdrawal (benzodiazepines are generally used for this purpose, as they have been for decades). Neither does it create a substitute, more socially acceptable addiction, as methadone does for heroin. Some say that it works by diminishing the physiological reward created by endorphins released by alcohol and thus eventually 'extinguishing' the conditioned response of continuing to drink to maintain the buzz (Sinclair, 2001); others simply state that it diminishes both the actual drinking and the 'craving' (O'Malley et al., 1992), without any theorizing, behaviorist or neurological. Either way, naltrexone, we are told, acts invisibly and automatically to readjust the level of craving, thus producing civilized conduct.

It is not coincidental, in terms of the dream of turning alcoholics into civilized beings who may on occasion have a drink, but won't even feel like drinking too much, that naltrexone is thought to target one kind of receptor in the brain rather than acting systemically. Like other fashionable 'smart' drugs (Prozac and Viagra being the most popular examples), the idea is that instead of acting systemically on the whole person or even the whole body, drugs are used to carefully select and affect one little invisible process in the brain in order to produce an adjustment – not a systemic 'cure' – in another specific, visible process. This cause-effect relationship, which is tightly bound to the Pavlovian 'extinction of conditioned response' paradigm by behaviorists, is also compatible with the more rationalist perspective favored by cognitive-behavioral psychology, a much more powerful paradigm in North American treatment circles than behaviorism.

More than a decade after being approved in the United States for alcoholism treatment, however, the magic bullet seems to have gained remarkably few converts. Dr Pearlman, a Boston physician who is typical of the hybridity of North American treatment approaches in that he has personal experience with Alcoholics Anonymous and runs a program combining cognitive-behavioral therapy and naltrexone, sees this drug as very promising for the large number of alcoholics who will not stay in AA or otherwise maintain total sobriety – but continues to promote total and lifelong abstinence as the preferred approach, a recommendation that reproduces the older, pre-smart drug figure of the born alcoholic characterized by a total identity (personal communication, 2 April 2001).

Dr Pearlman's ambivalence about fully embracing the smart-drug approach speaks to the tremendous influence of the Alcoholics Anonymous model upon North American alcoholism treatment (see Valverde, 1998, ch. 5). Naltrexone-based treatment allows alcoholics to still drink, and since AA has long fought against the introduction of 'controlled drinking' cognitive behavioral programs in North American treatment centers, naltrexone is likely to be tarred with the same brush. Even worse from the AA point of view, the treatment can only work if the person drinks at least some of the time, since otherwise the habit of drinking too much will not be acted upon at all, and hence not modified. (This was not clear at the time of the ReVia product announcement, but quickly emerged as an issue in treatment practice).

This illustrates a larger dilemma in governing conduct, and more specifically governing consumption and desire. As is well known, AA's key belief is that alcoholics are a distinct population with a distinct identity whose main feature is the inability to drink at all. It could, of course, be said that AA, like naltrexone, targets alcoholism rather than alcohol *per se* (since AA does not promote tighter liquor laws for all): but it does so by *governing through persons* – alcoholics have to self-identify as such. To go to AA, one has to believe, however pragmatically, in the existence of alcoholics as a distinct identity.

To attend a naltrexone clinic, by contrast, may in some cases (in the United States, in particular) require a temporary allegiance to the concept of 'alcoholism': but alcoholism can be refigured as a site-specific imbalance of brain chemicals, in the same way that prescribing SSRIs such as Prozac instead of anti-depressants can be seen as encouraging people to think of themselves as a regular person with a limited problem of serotonin imbalances, rather than as a member of a distinct and deviant group. Technologies of normalization and pathologization, whether lay like AA or medical, usually deploy, and reproduce, persons, diseases, and abnormal identities. If the logic of normalization is giving way to the logic of 'adjustment' and 'correction' in the field of mood disorders, as Nikolas Rose suggests (Rose, 2003), this may be linked to the waning of identity-based governance more generally.[3] But to make such claims, it is necessary to look a little more closely at the prehistory of smart drugs aimed at willpower and desire.

## The alcoholic brain as a moving target

Alcoholism is an unusual condition. First, unlike many other forms of deviance, it has never been successfully medicalized (Valverde, 1998). Second, those projects that have tried (unsuccessfully for the most part) to medicalize alcoholism have never fully effected the kind of reductionist biologization that has flourished in otherwise similar fields, such as mood disorders. In other words, alcoholism has been a hybrid condition, partly moral and partly psychological and physiological; and even those biomedical authorities who have at various points tried to read it as primarily physiological have shied away from reducing it to any one process or site – in today's paradigm, a particular deficiency or excess of certain brain chemicals.

The American National Institute on Alcohol Abuse and Alcoholism has gone a fair way down the biological road in recent years. This may be due to the saturation of the public sphere with messages about illegal drugs, a process that has had much influence on the way that alcohol is discussed, in health circles at any rate. The drug–alcohol relation is now the reverse of what it once was: while in the late nineteenth century the use of opiates such as laudanum or heroin was thought of as a species of 'inebriety,' nowadays drinking is subordinate to the paradigm of drug addiction. In a widely distributed newsletter on pharmaceutical innovations in alcohol research, a section entitled 'Alcohol and the Brain' reads as follows:

> All brain functions, including addiction, involve communication among nerve cells (neurons) in the brain. Each of the brain's neurons connects with hundreds or thousands of adjacent neurons ... Determining the specific neurotransmitters and receptor subtypes that may be involved in the development and effects of alcoholism is the first step in developing medications to treat alcoholism.
>
> (NIAAA, 1996)

This kind of discourse is typical of the immensely powerful NIAAA.

In keeping with the 'drugs and the brain' approach, then, alcoholism has come to be read more biologically than in the past in many definitions and programmatic statements. But whereas both legal drugs such as SSRIs and illegal drugs like cocaine and Ecstasy prompt drug companies and prevention professionals to produce and disseminate appealing, brightly colored pictures of brains and neurons (http://www.clubdrugs.org), alcohol research and alcohol education are not in fact as brain-centered as the 'alcohol and the brain' statement might suggest.

As an example, the pictures illustrating the educational posters and leaflets on alcohol distributed by Canada's foremost authority on the subject, the Center for Addiction and Mental Health in Toronto (formerly the Alcoholism Research Foundation of Ontario), show cartoon *people* – not brains, and not neurons – in various stages of disrepair.

Falling into gutters, getting ulcers, losing jobs, becoming disheveled, these dysfunctional characters, who look like they are drawn by the same artists who give us Dagwood and Doonesbury, give the impression that alcohol affects one's whole life, and that drinking is not a matter of a single brain chemical but a question of the self. The cartoons featuring stereotyped drunks/alcoholics are clearly not meant as scientific discourse, of course, but the fact that they are used by a prestigious scientific organization tells us something about the difficulties of moving fully into the era of smart drugs, moving beyond governing through identities. These difficulties are also visible in the strictly scientific discussion of the continued failure of science to locate any specific place or process within the human brain that might be to alcoholism treatment what serotonin receptors are to mood disorder treatment.

Along these lines, one of the key promoters of naltrexone treatment in the United States, Dr Joseph Volpicelli, openly admits that the quest for an actual, identifiable target for alcoholism treatments might be a wild goose chase: 'In contrast to opiates, alcohol does not seem to have specific agonist effects in stimulating opiate receptors directly. Rather, it is believed that alcohol has *nonspecific* effects on the lipid components of cellular membranes ...' (Berg, Volpicelli et al., 1990, p. 139). Other, less invested researchers have commented that the quest for a single biochemical process that could be targeted in order to moderate excessive drinking has no evidence to back it up, and has been motivated by nothing but 'the current Zeitgeist':

> the search for the neurobiological substrates of alcohol's effects on behavior in general and its positive reinforcing effects in particular now span several decades ... The goal of each of these lines of research has been to determine the involvement of a particular neurotransmitter in regulating this behavior ... Over the past two decades, a body of data emerged which clearly implicated several neural systems in the mediation of ethanol intake in different ways and different levels of specificity ... It is unlikely that any one single neuronal system would be identified as the primary-unitary system underlying ethanol's effects on behavior in general and ethanol intake in particular. What is far more likely is the emergence of a body of data pointing at a complex multi-system interaction in the mediation of ethanol's behavioral effects.
>
> (Amit and Smith, 1990, p. 161)

Even the very researchers who have the most to gain from proclaiming that they have found the appropriate target for anti-alcoholism magic bullets thus admit that such a target may not even exist. Nevertheless, the tremendous cultural capital acquired by genetics and neuroscience in recent years cannot but exercise an influence on the rather low-status and poorly funded pursuit of alcoholism treatment research. The quest for a pharmaceutical magic bullet for alcoholism will likely continue, therefore, hampered, but not fatally injured by the failure to 'discover' a specific biochemical process that can be isolated and manipulated – as recent-past history suggests.

In the 1970s, some researchers thought that lithium could be effective not only for the significant number of alcoholics who suffered from bipolar disorder but also for 'regular' alcoholics. One of them was the same Dr Sinclair who later developed the most internationally influential, proprietary naltrexone method, ContrAl (Kline and Cooper, 1975; Sinclair, 1975).

Along similar lines, in the 1980s, alcoholism researchers again attempted to find new uses for psychotropic drugs being developed for other conditions, mainly serotonin uptake inhibitors (SUIs). It was these 1980s experiments and trials with SUIs that first broached the possibility that people might be able to continue drinking – rather than needing to remain wholly abstinent – but without getting the pleasurable reinforcement thought to be the immediate, psychobiological cause of excessive drinking. A review of SUI-alcohol clinical studies concluded that some decrease in drinking was observed among subjects on SUIs. The Toronto authors who were leaders in this particular field concluded that SUIs '*may* act by decreasing desire to drink and the reinforcing properties of alcohol' (Naranjo and Bremner, 1990, p. 105; see also Naranjo, Bremner and Poulos, 1990).

That the effect of anti-addiction drugs on desire may have been broader than would be regarded as desirable by most patients, if not most clinicians, was only hinted at in these studies: Naranjo and Bremner reported some weight loss among their subjects even in very short (2–4 weeks) trials, but did not highlight this. Other, less invested researchers concluded, however, that SUIs produce a 'potent anorexic effect' (Amit and Smith, 1990, p. 167) that is anything but specific. This may have contributed to the fact that no SUI was (to my knowledge) approved by governments for use as an anti-alcoholism treatment. Similarly, an interesting finding that was not followed up, but that would surface again in the naltrexone trials was that SUIs were reported 'to have an effect on taste factors independent of the caloric value of the food' (Amit and Smith, 1990, p. 168). In other words, SUIs might work to decrease alcohol consumption only because they make both food and drink taste funny.

Other avenues for biochemically readjusting either the intake of alcohol or the desire to drink or both were explored during the 1980s and early

1990s. Generally, it was found that many psychopharmaceuticals do have a measurable effect on drinking, and studies that attempted to quantify desire rather than (or as well as) measure the drinking behavior also concluded that levels of what some social psychologists call 'craving' could be chemically manipulated to some extent.

## 'Craving research' and 'craving' research

In (a small number of) labs around the world, animal and human studies are still being conducted to discover which particular neurotransmitters are involved in the behavior of continuing to drink past the point of intoxication. But it is interesting that even those alcoholism researchers who have spent many years trying to get lab rats addicted to alcohol,[4] and who would thus be likely to favor biological explanations and solutions, do not want to question the old AA-influenced consensus about the need for psychotherapy.

A key figure promoting naltexone in North America is Dr Volpicelli, a professor at the University of Pennsylvania and head of a (private) treatment program called 'Assisted Recovery'. The very name of the program cleverly articulates naltrexone with the AA tradition. The use of the non-medical term 'recovery' mobilizes AA's belief that alcoholics cannot be cured or even treated, but can only be 'in recovery,' and that the lifelong state of being in recovery can only be achieved and maintained with the assistance of an entity that compensates for their deficiency, the 'Higher Power'. (The first of the 12 AA steps has the self-labeled alcoholic admit that he/she is powerless in the face of alcohol and that he/she can only be in recovery with the help of a Higher Power, an entity that may or may not be God, depending on the interpretation.) In Dr Volpicelli's clinics, naltrexone is presented exactly as the Higher Power is presented in AA literature: 'naltrexone is a tool to assist an individual in the process of recovery, and is not a cure in itself' (www.assistedrecovery.com). Whether naltrexone is a chemical substitute for the Higher Power or merely a physical supplement to the Higher Power is unclear, at least from the available literature, but either way the concept of 'being in recovery' clearly mobilizes the AA idea that being an alcoholic is a distinct and lifelong identity.

In the chain of naltrexone clinics owned and operated by ContrAl in Finland and other countries, and also in some sites in North America unconnected to Dr Volpicelli, the term 'recovery' is not mentioned. Along more rationalist lines, cognitive-behavioral psychologists devise programs designed to help heavy drinkers to acquire 'coping skills,' and naltrexone pills are given only to those who are willing to also attend 'coping' therapy sessions. Coping therapy avoids AA absolutism, and in general the traditional moralism about falling off the wagon. Instead, clients are given specific practical instructions about how to reorganize their time and space to minimize temptation, without insisting that abstinence is the one and only goal (personal communication). But even in these explicitly secular clinics, naltrexone is not used by itself the way that Prozac is often dispensed instead of therapy; naltrexone, in these clinics, is still a supplement – perhaps not for the Higher Power, but for the constant advice of trained harm-reduction specialists.

One of the two main American scientists engaged in testing and promoting naltrexone treatment, Yale's Stephanie O'Malley, explicitly describes naltrexone as a 'pharmacological support' for cognitive-behavioral therapy (O'Malley et al., 1992). This is also the way in which Dr Volpicelli's method is promoted. But while short courses of cognitive behavioral therapy, which are all the rage in relation to drugs, may be becoming more popular in relation to alcohol than the lifelong program promoted by AA, the North American dogma that alcoholism is a psychic problem and requires at least some therapy is never questioned. It is very significant that Stephanie O'Malley's seminal study, one of the two cited by government officials as justifying approval for this use of naltrexone, didn't even include a group of clients who were receiving the drug but no therapy. The whole study was designed to see whether naltrexone worked better with supportive (abstinence-oriented) therapy or with coping therapy (O'Malley et al., 1992; see also O'Malley et al., 1990). Similarly, attempting to hold at bay the specter that has haunted North American alcohol treatment for decades – the notion that alcoholics might be able to drink socially in a controlled manner – Berg and Volpicelli timorously wrote that 'while we certainly do not want to suggest that naltrexone could allow an alcohol-dependent individual to drink socially, the evidence presented here does suggest that treatment with naltrexone can give patients a "second chance" after a slip from abstinence' (Berg et al., 1990, p. 145).

On his part, the director of the NIAAA, while swept up in the magic-bullet biochemical enthusiasm immediately following government approval for naltrexone, nevertheless felt obliged to add: 'At the present time clinical research indicates that the best treatment results are achieved with a combination of pharmacotherapy and skilled counseling' (NIAAA newsletter Alcohol Alert, 33, 1996). That this was driven by dogma rather than research is clear from the fact that the NIAAA had to go out of its way to ignore O'Malley's Yale finding that naltrexone plus abstinence-oriented psychotherapy was not a good combination at all, and was indeed worse than being on placebo, a finding replicated in several other studies (see Agosti, 1994; Heinala et al., 2001; Sinclair, 2001).

The insistence on psychotherapy, particularly in the face of evidence that suggests that at least some kinds of therapy are actually contraindicated if one is on naltrexone, sits ill with any effort to reconceptualize excessive drinking as a merely physical matter of brain chemicals out of balance. If naltrexone, acting like the proverbial brain-science key in the lock of the opioid receptors, blocks the pleasure and/or the reinforcement effect of alcohol without any voluntary or cognitive activity on the subject's part, why is a pill not enough?

Dr Sinclair, a rat psychologist working at the Helsinki alcohol research unit that was until recently an integral part of the Finnish alcohol monopoly ALKO, is undoubtedly the most biologistic of all leading naltrexone researchers. He is a strict Pavlovian in a world that, especially in North America, is characterized by hybrid knowledges of the will and of desire. Even he, however, shies away from stating that taking a naltrexone pill when one anticipates being in a situation with a high risk of excessive drinking amounts to a cure. He believes that abstinence-oriented therapy is worse than useless for people on naltrexone – since if they are not actually drinking, the pill doesn't have an opportunity to adjust the neural pathways so as to wear away the conditioned response of prolonged drinking. But, instead of concluding that therapy is not particularly relevant, he nevertheless argues that 'coping' therapy should always be provided along with the pills (personal communication). The private clinics, mostly in Finland but also now as far away as Israel and Venezuela, using Sinclair's naltrexone method provide eight sessions with a psychologist or psychiatrist along with the pills, and Dr Pearlman's clinic in Boston provides five sessions. In Toronto, the Addiction Research Foundation's experimental naltrexone treatment program also makes naltrexone pills contingent upon attending therapy.

The unease about using pills and nothing else is not driven by scientific considerations, but rather by the lingering weight of the traditional American model of alcoholism, with its central belief in an alcoholic identity or personality. As deviant identities from nymphomania to homosexuality have been depathologized or otherwise reconfigured, the alcoholic is still an extremely successful and still fully deviant identity. Harm reduction and risk management are certainly present in the broader field of the governance of alcohol consumption – drinking and driving campaigns, most famously, stigmatize the person who doesn't give up the car keys after having had a drink, not drinking *per se*. But harm reduction perspectives have been very marginal within treatment circles, particularly in North America. That the subpopulation identified as alcoholics cannot learn to evaluate and manage their risks continues to be the great dogma of alcoholism treatment (especially in North America and in the Nordic countries) – despite the fact that it is a self-evident, circular truth, since alcoholics are defined precisely as those individuals who cannot manage the risks of drinking.

A certain poetic justice flows from this circular foundation: namely that if while on the naltrexone people classified as 'true alcoholics' (Jellinek, 1960) become able to drink socially rather than having to follow the old advice of targeting their willpower exclusively on that first tempting drink, this may open the door for the eventual 'extinction' – as the behaviorists would put it – of the whole field of alcoholism. Those formerly defined as alcoholics could be redefined as suffering from a minor chemical deficiency whose nature doesn't have to be elucidated before it is remedied by the ingestion of a pill that lowers one's craving/desire for the next drink. How this happens does not have to be known. AA has always advertised itself as effective 'because it works,' not because it has an ontologically coherent theory of alcoholism. And if naltrexone can prove that it works but at a lower emotional and psychic cost (because one doesn't have to actively refuse *all* drinks), then it will be regarded as more effective.

But while succeeding in targeting uncontrolled drinking as opposed to drinking in general, there are indications that it doesn't manage to target alcoholism quite as smartly as one might wish. Although evidence is scarce, and independent, disinterested evidence even scarcer, it seems that using naltrexone regularly may lower one's enjoyment of and/or interest in activities such as taking exercise, having sex, and eating sweets. Sinclair, who initially suggested one naltrexone pill every day, quickly changed his preferred protocol:

> Most earlier clinical trials have prescribed taking the antagonist every day. Selective extinction, however, is achieved by having patients take it only when they are drinking. In the optimal embodiment of this feature, patients avoid making other responses that are probably reinforced through the opioid system (e.g. eating highly palatable foods, having sex, jogging) while they are on the antagonist. Then when the craving for alcohol is manageable, they have days when no antagonist is taken, no alcohol is drunk, but these behaviors are now enjoyed. The simpler form of taking naltrexone only when [actually, before] drinking has also been called 'targeted medication'.
>
> (Sinclair, 2001, p. 7)

The language of 'targeting' is here deployed rather misleadingly: taking the pill only on days when one anticipates drinking can of course be described as targeting, but the reason for limiting the pills only to certain days is precisely that naltrexone may be acting not in a targeted manner but more generally and systemically on pleasure-seeking behavior. If one wants to target excessive drinking rather than pleasure more generally, then the patients will have to be told that they must drink at least some days.

Clinicians dispensing naltrexone have neither proved or disproved this generalizing, non-targeted effect. The only study I have found that inquired about sex drive reported that naltrexone does *not* lower men's sexual desire (Anton et al., 1999). But when pressed, one clinician told me about a patient who spontaneously reported that when playing squash while on naltrexone, he suddenly felt as if it were too much trouble to go and hit the ball, 'as if, "what's the point of squash anyway?" '. This is only one person's story; but it is perhaps significant that in at least one version of the brain picture used in ContrAl PowerPoint presentations, Dr Sinclair has 'chocolate' marked on one of the icons that are supposed to stand for alcohol-involved receptors. 'Gambling' is also written beside another icon, although it is unclear whether this indicates a separate receptor or another recently discovered function of the obviously multi-talented so-called opioid receptors.[5] Sinclair writes as if the broader effects that may or may not accompany the lowering of the urge to drink are self-evidently desirable:

> Pharmacological extinction [of the 'drink more' response] is a new form of medicine for treating behaviors that have become too powerful. What other behaviors might be treated? It can stop a rat from taking sweets. Can it control compulsive eating in humans? It can stop a rat from drinking methadone. Can it be used successfully to terminate methadone substitution therapy? Can it extinguish addiction to gambling, sexual obsession, compulsive thrill-seeking criminal activity? We will see.
>
> (Sinclair, 1998, p. 411)

These tentative findings about the drug are rather contradictory. Does medical science really want a drug that acts simultaneously on so many different sources of pleasure? The more powerful the drug, the worse the prospects will be for the long-sought project to target either the *behavior* of getting drunk or the *desire* to keep drinking (or both). Can targeted governance, in the realm of desire, ever be achieved? Will naltrexone pills enable alcoholics to drink socially, by contrast with the old-fashioned prohibitory strategy of AA – but without lowering their animal spirits and making them feel 'what's the point of squash anyway?' If drugs designed to curb excessive drinking and/or the desire to drink more end up affecting squash-playing or stock market gambling, who will decide what is and is not a 'side-effect'?

## Historical parenthesis on the free will: from maximizing willpower to reducing craving

From the private inebriate asylums for gentlemen of the late nineteenth century to the AA group meetings proliferating from the mid-twentieth century, alcoholism/inebriety treatments have generally sought to act upon the will. Alcoholism has never (except perhaps in France) been characterized as a purely physical disturbance – the ingestion of too much alcohol. It has been and still is regarded as a problematic relation of the self to consumption. How do you feel about your drinking? Do you drink to get away from problems? Can you stop drinking if you decide to stop? Do you hide your drinking from others? These are the kinds of questions one finds in risk assessments for 'alcohol dependence'. There is a consensus among experts that the existence of alcohol *problems*, even problems that are directly linked to drinking, like cirrhosis of the liver, is not a sure sign of the existence of alcoholism or alcohol dependence. Although the gap between alcohol problems and alcoholism is undoubtedly greatest in the United States, even in Europe alcohol dependence, like other forms of dependence, is not defined through physical symptoms or through objective risk factors. The World Health Organization concluded in 1989 that 'dependence' is 'a cluster of physiological, behavioral, and cognitive phenomena in which the use of a drug takes on a much higher priority for a given individual than other behaviors that once had higher value. A central descriptive characteristic of the dependence syndrome is the desire (often strong, sometimes overpowering) to take drugs, alcohol, or tobacco' (quoted in Lindstrom, 1992, p. 58).

Alcoholism is still characterized, then, not as an imbalance of fluids, but rather as the existence of a powerful inner compulsion. 'Loss of control,' the term popularized by alcohol science in the United States in the 1960s, is still found in alcoholism diagnosis and treatment literature, despite its transparently puritanical overtones.

Traditionally, it was thought that the road back to control lay in strengthening and maximizing the drinker's willpower. AA tells people to put all of their willpower efforts in a single basket, concentrating on saying 'no' to the first drink. Targeted willpower, if you like. The classical AA definition of success is often clouded, today, by the discourse of self-help, self-esteem, dysfunctional families, and the rest. But in its classic form, AA provides a modest and yet powerful targeted tool with excellent audit potential: as long as you are not drinking (and are going to meetings), you are sober. You'll never reform or get cured; there is no normalization possible for the alcoholic. But you can lead a normal life if you simply focus all your willpower – and that provided by the Higher Power – on a single act: refusing to take the first drink.

Other, more rationalistic approaches try to modulate the drinker's willpower so as to regulate rather than prevent drinking. This more ambitious project is enacted through such harm reduction techniques as learning to space one's drinks farther apart or readjusting the context within which drinks are chosen. People are encouraged to develop strategies such as purchasing attractive substitutes or modifying their daily time and space routines. People following these 'coping' programmes have to exercise their will in relation to numerous acts and situations, but the cognitive faculties also get quite a work out: the drinker is in a constant state of self-monitoring.

There are thus huge differences between AA and the techniques taught by more rationally oriented cognitive-behavioral programmes aimed at regulating drinking; but what they have in common is that they act on the drinker's capacity and willingness to say 'no' to either all or only to some drinks. In cognitive-behavioral therapy, the language of the will has disappeared, replaced by the language of 'coping,' 'situations,' 'decisions,' and 'contract'. But the ghost of the free will is clearly visible: the goal of therapy is to make it easier to say no to 'bad' things. In keeping with the traditional quasi-religious goal of strengthening one's capacity to say no to temptation, cognitive programmes follow a logic of maximizing personal capacities. The opiate antagonists, however, are spoken about not in the nineteenth-century language of boosting willpower, but rather in the 1970s language of budget slashing and deficit reduction – more specifically, craving reduction. Semantics do matter: it would be as consistent with the scientific evidence to claim that naltrexone is a Popeye-spinach of the will, or of the rational capacity, as to say that it reduces craving. So if we are to understand what underlies the language of 'craving,' we must look elsewhere than to objective or scientific factors.

The clinical psychology literature on 'craving' is full of semantic inconsistencies. The word 'desire' is sometimes used as a synonym for craving, as in the WHO definition of dependence cited above. At other times, however, 'craving' is defined as a time-specific irresistible urge for something specific – as in the cliché of pregnant women craving pickles and ice cream.

In this second usage, 'craving' is *targeted desire*. But whether craving is thought of as always there, like psychoanalytic desire, or as a momentary urge, researchers in this area cannot seem to get to first base. Can craving be measured, inside or outside the laboratory? Is craving purely subjective? If it can't be measured and scientifically operationalized, should the concept of craving be retained at all? These fundamental questions recur in virtually every study.

Of course ontological doubts are hardly unique to this field. But a perhaps more serious difficulty is that there is no agreement as to how to measure desire/craving. An ingenious tool developed by smoking-cessation researchers, the Ecological Momentary Assessment mini-computer, seems promising for measuring craving *in vivo*: this is a Palm Pilot programmed to beep every so often to remind the subject to input a few measures recording mood and desire (Stone and Shiffman, 1994). But for whatever reason this machine for recording desire in real time and in real life has not been used by alcohol and drug researchers.

Researchers have discovered one thing, which is that craving is not coterminous with Pavlov-type physical symptoms such as salivation, since, like hunger, it can occur at any time, and is not necessarily linked to physiological processes. More significant for our purposes is the finding that craving, as measured in self-report scales, is not always associated with quantity of drinking (Tiffany and Conklin, 2000, p. 148). Finding that not all craving occasions lead to drinking would have led earlier scientists to reflect in laudatory terms about the strength of human willpower; but, in keeping with current orthodoxy, the authors of this study merely state that 'the regulation of drug use in experienced addicts can function independently of the processes that control craving' (ibid.).

The craving researchers have also discovered that craving, far from being an irrational urge, is strongly influenced by cognitive factors. Experimental subjects who know that they will not get actual drugs or alcohol in the course of a study have a tendency to exhibit far fewer physical symptoms of 'craving' and to report far less psychological craving (Meyer, 2000, p. 222). So which brain process contains 'craving'? Nobody knows.

Wanting to avoid these difficulties, some of the people in the field argue that 'craving' suggests biological need and should thus be abandoned, since the whole problem of drugs and alcohol is that, unlike food, we don't only take them when we feel a biological urge to consume. So what about pleasure? Is 'pleasure' a useful term? Or is 'pleasure' simply the lay term for the satisfaction and reinforcement of a craving (Volpicelli et al., 1995; Abrams, 2000)? And so forth.

It is thus clear that although we may now be past the age of 'diseases of the will' and into the age of deficits of brain chemicals, it is nevertheless proving very difficult to operationalize alcoholism in brain terms. Shifting the focus from the ever-elusive will to MRI pictures of the brain has not eliminated the chronic ontological ambiguities that have characterized 'alcoholism' since its inception in the late eighteenth century, and that seem to plague psychological research on desire and willpower more generally. There is a consensus that drinking is affected by a host of biochemical (and environmental) processes rather than located on a single biochemical site. To confound targeted governance further, the allegedly smart opiate receptor antagonists are not so smart after all – or are perhaps too smart. Naltrexone also works to alleviate heroin and even methadone withdrawal and perhaps even to diminish compulsive gambling.

Sexual desire seems to have been targeted more successfully – if the runaway success of Viagra is any indication (which it may not be, but that would be another chapter). By contrast, the immoderate desire to drink remains a moving target.

## A final note on the temporality of 'smart' governance

The targeted or 'smart' governance of conduct could be seen as a modest, less invasive governance than that favored by the Fabian and welfare-state era dreamers. Never mind grand reform: identify the little problem, the broken window of the soul or of the house, fix it before it gets big, and get on with life – that seems to be the advice we get at the therapist's office as well as at the police department. However, if targeting is to succeed, and if smart bombs are ever to really become smart, assessments of the results will have to be constantly fed back to the authorities. Targeting is rarely right on from the beginning, and even if it works, it can always be improved. Constant assessments and audits are thus necessary for the continuation of 'smart' targeted governance. This hardly amounts to governing less.

What about naltrexone? Can alcoholics take the pill for a while to ensure the 'extinction' of the conditioned response and then consider themselves cured? Or is the targeted intervention a lifelong sentence?

The inventor of the 'Sinclair method' tells us that although naltrexone should not be taken every day (since it may affect other pleasures), it could or should be taken for life: 'There is no experimental or theoretical justification for terminating naltrexone treatment after a fixed period of time of any duration. The only real justification is financial: taking naltrexone every day for the rest of one's life would be expensive' (Sinclair, 2001, p. 8).

And what Dr Sinclair does not say is that anyone taking naltrexone will have to be constantly monitoring the effects of the drug on other pleasures, adjusting the dose accordingly.

Thus, even the best targeted governance requires assessment and adjustment; but in relation to drinking, no matter how targeted the medication, the fundamental problem of alcoholism treatment in the age of 'brain science' is that no alcohol receptor has ever been located in the brain. The only part of the brain with any claim to having a specific link with alcohol is one that has a documented relation with opium and an anecdotal relation with chocolate eating and sport participation: therefore, the 'smart drug' approach is unlikely to ever succeed.

That drinking conduct may thus be a fascinating area for research, precisely because 'smart drugs' are highly unlikely to ever be found, is not something that many experts in the treatment field are likely to say, in their grant applications at any rate – but it is a thought of some interest to non-specialists in the broader fields of the human sciences.

## Notes

* This is an updated and revised version of a paper given at the 'Risk and Morality' conference and subsequently published in the resulting collection, *Risk and Morality*, ed. Richard Ericson. My thanks to Nikolas Rose for sharing his contribution to that collection with me in advance and engaging in conversations that were important both to the conference paper and to this chapter.

1 Performance assessments and audits (Power, 1994) are other knowledge practices that fit well with 'smart' or targeted governance, and are actually necessary if the targeting process is to be refined and improved, but for present purposes I will focus mainly on the moment of targeting, not the subsequent moment of evaluation, assessment, audit, and refinement.

2 In the 1950s the World Health Organization organized a debate about whether alcohol ought to be listed as a 'habit forming drug' or an 'addictive' drug. This was no mere academic exercise, since drugs classified by WHO as 'addictive' are supposed to be regulated tightly according to international standards, something that no wine-producing nation would ever contemplate. In the end WHO created a third, compromise category, in-between 'habit forming' and 'addictive' (Valverde, 1998, pp. 39–40).

3 Elsewhere I explore the waning of the 'homosexual' identity – see Valverde, 2003, chs 4 and 5.

4 Alcoholism research is plagued by the fact (almost never mentioned in the literature) that it is very difficult to get lab rats to drink a great deal of alcohol – they seem to drink a bit but then stop. In the 1960s the Alcoholism Foundation of Ontario used real alcohol and real people for their research, or so local Toronto anecdote about a bar set up in the basement of the Foundation suggests; but needless to say no ethics review board in North America today would approve a protocol involving plying alcoholics with drink. Rats' lack of interest in getting drunk thus remains a problem.

5   It is beyond the scope of this paper to study how the 'opioid' receptors were discovered and named; for present purposes it suffices to point out that these neurological entities remain associated with opium consumption by virtue of their name even though they have been shown to be active when all manner of stimulating activity is occurring. That is, they could just as well be called 'pleasure receptors'.

## References

Abrams, D., 'Transdisciplinary Concepts and Measures of Craving: Commentary and Future Directions', *Addiction*, 95 (Supplement 2, 2000): 237–46.

Agosti, V., 'The efficacy of controlled trials of alcohol misuse treatments in maintaining abstinence: A meta-analysis', *The International Journal of the Addictions*, 29, 6 (1994): 759–69.

Amit, Z. and B. R. Smith, 'Neurotransmitters regulating alcohol intake', in C. Naranjo and E. Sellers (eds), *Novel Pharmacological Interventions for Alcoholism* (New York: Springer Verlag, 1990), pp. 161–83.

Anton, R. F., D. Moak, L. R. Waid, P. K. Latham et al., 'Naltrexone and Cognitive Behavioural Therapy for the Treatment of Outpatient Alcoholics: Results of a Placebo-Controlled Trial', *American Journal of Psychiatry*, 156 (1999): 1758–64.

Berg, J. B., J. Volpicelli, D. Herman and C. O'Brien, 'The Relationship of Alcohol Drinking and Endogenous Opioids: The Opioid Compensation Hypothesis', in C. Naranjo and E. Sellers (eds), *Novel Pharmacological Interventions for Alcoholism* (New York: Springer Verlag, 1990), pp. 137–47.

Foucault, M., *The Order of Things: An Archeology of the Human Sciences* (New York: Vintage, 1973).

Heinala, P., H. Alho, K. Kiianmaa, J. Lonnqvist, K. Kuoppasalmi and J. D. Sinclair, 'Targeted use of naltrexone without prior detoxification in the treatment of alcohol dependence: A factorial double-blind, placebo-controlled trial', *Journal of Clinical Psychopharmacology*, 21, 3 (2001): 287–92.

Jellinek, E., *The Disease Concept of Alcoholism* (New Haven, CT: Hillhouse Press, 1960).

Kline, N. S. and T. B. Cooper, 'Evaluation of Lithium Therapy in Alcoholism', *The Effects of Centrally Active Drugs on Voluntary Alcohol Consumption* (Helsinki, 1975).

Lindstrom, L., *Managing Alcoholism: Matching Treatments to Clients* (New York and Oxford: Oxford University Press, 1992).

Meyer, R., 'Craving: What can be done to bring the insights of neuroscience, behavioral science and clinical science into synchrony', *Addiction*, 95 (Supplement 2, 2000): 219–27.

Morrison, J., 'The Government-Voluntary Sector Compacts: Governance, governmentality, and civil society', *Journal of Law and Society*, 27, 1 (2000): 98–132.

Naranjo, C. and K. E. Bremner (1990), 'Evaluation of the effects of serotonin uptake inhibitors in alcoholics: A review', in C. Naranjo and E. Sellers (eds), *Novel Pharmacological Interventions for Alcoholism* (New York: Springer Verlag, 1990), pp. 105–20.

Naranjo, C., K. E. Bremner and C. X. Poulos, 'A human model for testing drug-induced concomitant variations in alcohol consumption and the desire to drink', *Novel Pharmacological Interventions for Alcoholism* (New York: Springer Verlag, 1990), pp. 288–98.

NIAAA, *Press Release: 'Naltrexone approved for alcoholism treatment'*, 17 January (1995).

NIAAA *Alcohol Alert*, 3 July (1996).

NIDA, *News Release: 'Alcohol researchers identify new medication that lessens relapse risk [nalmefene]'*, 30 August (1999).

O'Malley, S. et al., 'Naltrexone in the Treatment of Alcohol Dependence: Preliminary findings', in C. Naranjo and E. Sellers (eds), *Novel Pharmacological Interventions for Alcoholism* (New York: Springer Verlag, 1990), pp. 148–59.

O'Malley, S., A. Jaffe, G. Chang, R. Schottenfeld, R. Meyer and B. Rounsaville, 'Naltrexone and coping skills therapy for alcohol dependence: a controlled study', *Archives of General Psychiatry*, 49 (1992): 881–8.

Power, M., 'The Audit Society', in A. Hopwood and P. Miller (eds), *Accounting as a Social and Institutional Practice* (Cambridge: Cambridge University Press, 1994), pp. 299–316.

Rose, N., 'The Neurochemical Self and its Anomalies', in R. Ericson (ed.), *Risk and Morality* (Toronto: University of Toronto Press, 2003), pp. 407–37.

Sinclair, J. D., 'The effects of lithium on voluntary alcohol consumption by rats', *The Effects of Centrally Active Drugs on Voluntary Alcohol Consumption* (Helsinki, 1975).

Sinclair, J. D., 'New treatment options for substance abuse from a public health point of view', *Ann Med. Of the Finnish Medical Society*, 30 (1998): 406–11.

Sinclair, J. D., 'Evidence about the use of naltrexone and for different ways of using it in the treatment of alcoholism', *Alcohol and Alcoholism*, 36, 1 (2001): 2–10.

Stone, A. A. and S. Shiffman, 'Ecological Momentary Assessment [EMA]', *Annals of Behavioral Medicine*, 16 (1994): 199–203.

Tiffany, S. and C. Conklin, 'A cognitive processing model of alcohol craving and compulsive alcohol use', *Addiction*, 95 (Supplement 2, 2000): 145–53.

Valverde, M., *Diseases of the Will: Alcohol and the Dilemmas of Freedom* (Cambridge: Cambridge University Press, 1998).

Valverde, M., *Law's Dream of a Common Knowledge* (Princeton: Princeton University Press, 2003).

Valverde, M. and M. Mopas, 'Targeted security', in W. Larner and W. Walters (eds), *Global Governmentality. Governing International Spaces* (London: Routledge, 2003), pp. 233–50.

Volpicelli, J., A. Alterman, M. Hayashida and C. O'Brien, 'Naltrexone in the treatment of alcohol dependence', *Archives of General Psychiatry*, 49 (1992): 876–80.

Volpicelli, J., N. Watson, A. King, C. Sherman and C. O'Brien, 'Effect of naltrexone on alcohol "high" in alcoholics', *American Journal of Psychiatry*, 152 (1995): 613–15.

# Part IV
# Self and Morality

## Introduction

The following part is dedicated to alternative conceptions of will as suggested by concepts of relational responsibility and moral realism. Although philosophical in their outlook, the contributions are neither preoccupied with the well-discussed indeterminism/determinism nor the freedom/coercion antagonisms, but offer novel insights into the role of values and morality. In so doing, they adhere to a concept of free will compatible with neuro-cognitive findings, albeit in different ways. While Kenneth J. Gergen adopts a constructivist view by way of viewing agency and determinism as culturally evolved attributions, ultimately designed to coordinate our actions, Tillmann Vierkant advances a realist notion of autonomous action determined by a self that is far more than a subject: as an ideal type, the self encompasses a relationship between an organism and its natural and social environment. It is thereby able to accord its actions to the demands of inner and outer circumstances. Neither account dispenses with the concept of voluntary action. By contrast, both accounts insist on norms and values, such as responsibility for one's actions as discursive or evaluative resources guiding social practices enacted by individuals or groups.

In the framework of the argument advanced in this volume, agency, autonomy, and responsibility emerge in practices of attribution. The very capacity to attribute agency, autonomy, and responsibility (individually or collectively) is seen as social capital embedded in self-technologies (such as self-help, see Maasen, Sutter and Duttweiler, Chapter 1 in this volume), in which it exerts highly governmental functions. Namely, it is on the basis of such practices that a social capital named 'governing oneself and others' has emerged. Moreover, the emergence of this capital forces us to rethink acceptable modes of producing 'the social,' 'rational discourse,' and 'morality' in a highly individualized, if not fragmented world fraught with uncertainties.

In this situation, a 'postmodern ethics' seems to be called for. To be sure, a postmodern ethics would not be about a situation where 'anything goes,' but rather about acknowledging a world without recourse to a fixed set of guiding codes or principles. As Patrick Bracken and colleagues point out, 'postmodern thinkers would hold that by focusing on the responsibility to act, traditional ethics has had to "fix or close down parameters of thought and to ignore or homogenize at least some dimensions of specificity or difference among actors. To act in this sense means inevitably closing off sources of possible insight and treating people as alike for the purpose of making consistent and defensible decisions about alternative courses of action. The modern thinker associates the commitment to this sense of responsibility with self-justification either in the sense of moral-uprightness or pragmatic effectiveness" (White)' (Bracken, Giller and Summerfield, 1997, p. 436). Postmodern ethics, however, is based on ongoing dialogue, with others as well as with oneself.

It is within this type of morality that modern governmental technologies operate, be they self-help books, or participatory, mediating, or therapeutic practices: they all contribute to a morality consistent with the demands of neoliberal society. This morality is neither about avoiding moral wrongdoing so as to keep up an empty husk of 'shoulds' and 'oughts', nor is this morality fully characterized by exercising the virtues (justice, fortitude, prudence, and temperance). Rather, this is a morality of decision, commitment, responsibility, and an ethics of autonomous selfhood – all of which, as one may argue on the basis of the following two chapters, rest on our (culturally evolved and always negotiated) capacity to engage in reasoned interaction with others as well as with oneself. By doing so, we create the social and normative reality aspired to as well as the agents and their agencies, hence, the subjects constructing this very social and normative reality. The theoretical strength of the concept of governmentality lies precisely in the fact that it construes neoliberalism as a political project endeavoring to create a social reality that it suggests already exists.

Especially with regard to self-oriented technologies, it is commonly lamented that now 'we have selves whose absolute detachment from webs of interlocution brings about an incommensurability that makes moral discourse incoherent. We have a world of sole authors with their stories, whose basis is, perhaps a "feeling good" that is entirely private and cannot be fully and rationally explained to any other person. There is no shared understanding of what the Good is. There is only the agony of fragmentation' (Greenberg, 1994, p. 205). Although this line of critique does have its grain of truth, it fails to consider that self-technologies are social practices: whichever way we work on ourselves, even if in allegedly more serious ways (e.g., psychoanalytically or by ZEN-meditation), all interventions into one's self rely on techniques that ultimately do have their disciplining effects on ourselves as well as on our perspective toward the world around us – family, organizations, and so on. The analysis of governmental technologies reminds us that sociality ultimately rests on both a political and an ethical economy of the body. We can decipher a neoliberal governmentality in which not only the individual body, but also collective bodies and institutions (public administrations, universities, etc), corporations, and states have to be 'flexible,' 'autonomous,' and 'responsible,' requiring us and others to engage in pertinent activities.

In this perspective, the analytics of government focus on the systematization and 'rationalization' of a pragmatics of governance. The term rationality refers to specific socio-historical practices; it refers to specific and empirically observable social practices that can be analyzed with respect to 'how forms of rationality inscribe themselves in practices or systems of practices, and what role they play within them, because it's true that "practices" don't exist without a certain regime of rationality' (Foucault, 1991, p. 79). Practices of excessively attributing agency, autonomy, and responsibility are part of the neoliberal rationality of government, as are notions of individuals who determine themselves by way of achieving an ideal relationship between their organism and their social and natural environment.

Within this rationality, ethics and morality adhere to the demands of audit society: where, earlier, 'discipline sought to fabricate individuals whose capacities and forms of conduct were indelibly and permanently inscribed into the soul – in home, school, or factory – today control is continuous and integral to all activities and practices of existence.' As Nikolas Rose observes: 'we are required to be flexible, to be in continuous training, lifelong learning, perpetual assessment, continual incitement to buy, to improve oneself, constant monitoring of health, and never-ending risk management. In these circuits, the active citizen must engage in a constant work of modulation, adjustment, improvement, thereby responding to the changing requirements of the practices of his or her mode of everyday life' (Rose, p. 97 in this volume). This amounts to saying no less than the flip side of late-modern or postmodern ethics reveals an ethics of control, regardless of whether they are conjointly created (Gergen, Chapter 8) or explicitly in line with our self-model (Vierkant, Chapter 9). In this perspective, a relational ontology and reason-responsive self appear as most constitutive ingredients of the neoliberal moral economy.

# References

Bracken, P., J. E. Giller and D. Summerfield, 'Rethinking Mental Health Work with Survivors of Wartime Violence and Refugees', *Journal of Refugee Studies*, 10, 4 (1997): 431–42.

Foucault, M., 'Questions of Method', in G. Burchell, C. Gordon and P. Miller (eds), *The Foucault Effect: Studies in Governmentality* (Chicago: University of Chicago Press, 1991), pp. 73–86.

Greenberg, G., *The Self on the Shelf. Recovery Books and Good Life* (New York: State University, 1994).

White, S. K., *Political Theory and Postmodernism* (Cambridge: Cambridge University Press, 1991).

# 8

# From Voluntary to Relational
# Action: Responsibility in Question

*Kenneth J. Gergen*

Concern with the animating source of bodily movement has occupied human inquiry for centuries. What is the source of human action, and what is the nature of its departure during death? This was indeed the question posed by Aristotle in his writings about psychology and the will. As he saw it, there is an active force within the person that is responsible for bodily animation. To this force he assigned the concept of what is sometimes translated as 'soul'. To the soul is ascribed the 'power of producing both movement and rest' (p. 127). It is this originary font of action (arche) within the individual which, in the *Nicomachean Ethics*, Aristotle describes as the basis of willed or voluntary behavior. Aristotle's question remains ours today, as we ponder the question of voluntary agency. And, indeed, one may conclude that Aristotle's conclusion continues to be very much alive.

This concept of a psychological 'originator of motion' has undergone many transformations over the course of history. Importantly, these transformations can typically be traced to valued institutional or political ends. Thus, for the early Christian church (and Catholicism today) the origin of action lies within the soul. And as Foucault (1978) has described it, with the Church claiming the power of intercession from on high, the confessional was essentially a means of maintaining control over the polity. With the Enlightenment, the soul was largely replaced by human reason as the originator of action. Human agency thus moved from the provenance of the sacred to the secular, and with it the control over individual action became a matter of civil as opposed to religious governance. Police interrogation replaces the confessional, and the courts replace the judgment of God. It is in this context that one can appreciate Kant's writings that viewed the individual actor as both autonomous and responsible. The concept of the categorical imperative is essentially a means of eliding moral responsibility with the power of reason. In significant degree we may understand contemporary psychology, and its explorations of cognition, as a secular extension of the Enlightenment tradition. And, as Rose (1990) would have it, psychological science – and its clinical extension – now functions as a contemporary priesthood. The clinical interview, no less than the Catholic confessional, determines the adequacy of individual action.

Yet, the case of psychological science is also diagnostic of a major shift in cultural belief. Although psychologists will trace action to psychological origins, the concept of voluntary or free agency has largely been abandoned. This is primarily because psychological science allies itself with the modernist movement toward scientific explanation. With this shift toward science, explanations become largely mechanistic. Mechanistic explanation entails that for every action there is an antecedent cause, and cause–effect sequences find their origins in more fundamental laws of nature. Within this tradition, there is little place for an independent, originating force within. We may speak of God as an uncaused cause, but to attribute such a capacity to the human being would undermine the entire scientific project. To presume that individuals could decide at any moment to do as they willed, would render the scientific attempt to 'predict and control' human behavior defunct.

Thus, we find ourselves today with two major but competing traditions, each profoundly woven into both our major institutions and the rituals of everyday life. On the one hand is the tradition of voluntary agency, and on the other the causal determination of action. It is this antinomy that sets the stage for what follows. My attempt here is, first, to explore the implications of viewing both agency and determinism as cultural constructions. This will enable us to inquire into their consequences for cultural life. My particular concern is with the individualist ideology that is favored by these concepts and a family of consequences that many will find injurious to human wellbeing. Such consequences will be illustrated in practices of distributing responsibility. Then, given the problematic effects on cultural life, I shall open discussion on an alternative to the individualist ontology. My special interest here is in the plausibility of a relational ontology. With a relational understanding of human action sketched out, we are positioned to consider alternative and potentially more promising forms of cultural practice. We shall explore the potentials of relational as opposed to individual responsibility.

## Agency/determinism as discursive resources

In calling attention to the historical transformations in the concept of will or agency, we also make clear its socially constructed character. When we thus remove its ontological grounds and understand it as a cultural construction, we are also invited to consider the social functions of such discourse in contemporary society. By the same token, the discourse of causal determination may also be viewed historically. The earliest account of mechanistic cause, for example, can also be traced to Aristotle. Among the various forms of explanation outlined in Aristotle's *Physics*, were *final cause* (for many scholars encompassing explanation by personal agency), and *efficient cause* (represented in the mechanistic tradition). The use of causal discourse has also shifted over time, and today, for example is pivotal within the science of psychology but scarcely used in quantum physics. In effect, let us view both agentive and deterministic forms of explanation as discursive resources available for various uses in society. Inquiry is then invited into the forms of life that are sustained by these concepts.

Scientists and humanists have long waged battle over the use of these explanatory forms. On the one side, scientists have come under fierce attack because the use of deterministic explanation undermines the validity of voluntary agency. To do so removes the rationale underlying our traditions of moral responsibility, justice, and jurisprudence. Further, the picture of the person as a form of robot reduces the value placed on human life. If the person is simply a machine, then warfare and genocide are rendered more plausible as solutions to conflict. And, too, it is opined, a century of behavioral science has yielded very little in our improvement to 'predict and control' human behavior. In general, the best predictions of people's actions result from simply asking them about their intentions.

On the other side, behavioral scientists have despaired of using agentive explanations, both within the sciences and daily life. Attributing action to voluntary agency is a pseudo-explanation, it is said. It is simply to transcribe the description of an activity into an identical tendency within. If we say that the reason someone robbed a bank is because he chose to, we have done nothing more than say that he had an internal tendency to rob a bank. Most importantly for the scientist, such explanations discourage a search for other, more empirically available factors (environmental and hereditary) that may account for such activity. Or as Skinner (1971/2002) would have it, the concept of free agency is counter-productive; if problems of human conduct are to be controlled, it is essential to locate their material causes.

Responding to these accusations, those championing free agency argue that acceding to causal explanations invites us to view individuals as mere objects. In particular, we come to view individuals as 'manageable,' as entities that can be moved in one direction or another by those who have the resources to control the relevant conditions. It is this mentality, it is argued, that lends itself to command and control management practices, that suppress the opinions and needs of organizational participants, that supports top-down pedagogical practices and policies that disregard the motives and values of students, and that favor the control of crime and terrorism without understanding their meaning for the perpetrators. It is simultaneously a mentality that breeds suspicion of all those in positions of power – in government, business, law, and advertising. Regardless of their motives, a hermeneutic of suspicion prevails.

## Agency/determinism and human responsibility

There is no obvious resolution to these cultural battles. But in many respects such an impasse is to be welcomed. If we view determinism and agency as discursive resources, we should be pleased to see both remaining robust and simultaneously relativized. Both may have circumscribed utility, and for one to extinguish the other would be a cultural loss. At the same time, we should be sensitive to problems inherent in these yoked forms of explanation, and open to entertaining alternatives of additional promise. To argue that such concepts are contingently useful should not prevent us from inquiring into conditions under which they are not useful, or indeed, inimical to human well-being. And, should we locate shortcomings, we are invited to consider the development of alternative constructions. If we live and die as a result of the way in which we construct the world, then we should be ever watchful for conditions under which new constructions are required.

Although touching already on the problems inherent in both orientations to understanding, my particular concern in the present chapter is with the common practices of allocating responsibility. The presumption that individuals are responsible for their actions is, as we have seen, pivotal to the agentic view of human action. And this presumption is widely championed, not only because it is for many people quite accurate (people simply are moral agents), but because it acts as an important means of social control and achieving justice. We thus hold individuals responsible for callous behavior in daily relationships, for poor performance in the classroom, for shoddy work in the organization, and for murder everywhere in society. The result is typically some form of negative sanction, with the view of (1) meting out just punishment, (2) discouraging such behavior in the future, and (3) discouraging others from such untoward behavior. As we have also seen, the determinist alternative is typically discredited because it either excuses people for all wrong-doings, or simply ends in administering regimens of punishment (in the name of relearning) that fail to achieve a sense of justice.

I am not prepared to argue that practices of distributing responsibility make no contribution to an ordered society. However, there are considerable costs that accrue as a result:

- The individual is under continuous threat for possible shortcomings. Virtually all one's actions can be found wanting from at least one standpoint. Whether one is eating carbohydrates, smoking a cigarette, watching television, phoning while driving, crossing a street between intersections, or driving a large car, there is a significant group prepared to voice moral blame. The threat of being placed under judgment is unrelenting.
- A wedge is placed between the person under evaluation and the system of rules – either formal or informal – from which judgments about him/her are derived. It is *others* who are holding one to a standard, and the result is often the alienating of the individual from the standards. The standards are not embraced as one's own, but viewed as alien controls. They do not represent intrinsic goods, but are there to be subverted if one can. As many taxpayers feel, it is the government against my cunning; as many drivers on the highway feel, it is the police against my ability to speed without detection.
- The individualist tradition honors the right of individuals to make their own judgments of right and wrong. The result is an abiding sense that others have no right to make judgments of one's behavior. As liberal individualist philosophy suggests, each person has the right to act as he or she pleases, so long as it does not interfere with the actions of others. Thus, there is little reason to join in deliberation on the common good.
- Partly owing to the preceding, the common reaction to being judged is not contrition, but alienation and resentment. Even murderers often feel their actions were justified ('he was no good,' 'he was a threat,' 'he stupidly got in the way,' 'this is a way of life in my neighborhood'.) Prison is primarily used as punishment, as opposed to a correctional device. The result is seldom the production of chastened individuals but, rather, that of hardened and sophisticated criminals.
- By holding single individuals responsible for untoward actions, we are discouraged from exploring the web of relations in which the individual is enmeshed. We hold the individual terrorist to blame, for example, without taking into account the religious traditions in which the individual participates, nor the relationship of this tradition to the

larger culture that threatens it. In this sense, individual blame represents a vast simplification and suppression, a way of terminating deliberation on complex issues. Such deliberation might also reveal our own complicity in bringing about the act we wish to punish.

Again, I am not suggesting that because of their injurious consequences we abandon these discursive traditions. As we have seen, they do have contingent functions. However, their problems do invite us to consider the potential of alternative ontologies. If will and determinism are discursive achievements, then through further dialogue we may generate alternative constructions with different consequences. It is to one such alternative that I now turn.

## Toward a relational ontology

In a broad sense, both the concepts of agency and causality lend themselves to a picture of an atomized society. In both cases society is composed of individual, bounded units. Whether moved to action by will, or by an external causal force, the concept of individual units remains. There has been much recent interest in moving beyond this view of bounded being, and with it the agency/determinism binary, and articulating a relational account of human action. Such a view would replace the individual as the atom of society with relational process. While this is not an appropriate context for reviewing the emerging corpus of literature on this topic, it is fair to say that Wittgenstein's later writing (1953) has played a primary germinating role. It is not simply that such writings question the ontological status of all mental predicates (including the will), but more importantly, they locate the genesis of all ontologies within the linguistic practices of persons in relationship. The metaphor of the language game is pivotal here, as it suggests that all words come into meaning through their communal use. It is thus that we may reasonably view such terms as 'will' and 'determinism' as cultural resources. Or to extend the logic, the ontological presumption of individual or bounded selves is lodged within a relational practice. It is not individual agents who enter into relationships, but relational practices that give rise to the very discourse of the individual.

One important opening to this more radical conception of the relational emerges from contemporary discourse analysis. Such analysis typically focuses on the pragmatics of discourse use. In the case of mental discourse, then, the analyst directs attention to the way it functions within relationships. For example, in Potter and Wetherell's (1987) ground-breaking work, the concept of 'attitude' is shorn of mental referents, and as they see it, serves to index positional claims within social intercourse. An attitude, then, is essentially a social claim ('I feel ...,' 'My view is ...,' 'I prefer ...'), and not an external expression of an internal impulse. Similarly, Billig's (1996) treatment of reason focuses on the way in which people argue, thus defining reason not as a mental event but a relational process. In the same way, memory becomes a social or communal action, a way of talking or acting on certain occasions (Middleton and Brown, 2005).

If meaningful language comes into being through relational coordination, the same may be said for meaningful actions outside the verbal realm. In this sense, our gestures, gaze, and posture are fashioned within the matrix of relationship. And if we are to function adequately in society, the ways in which we walk, sit, or stand will be those deemed appropriate by standards negotiated in relationships.

The emotions furnish a convenient illustration of the relational perspective. Traditionally the emotions have been viewed as inherent properties of the individual, biologically based and evolutionarily grounded. In contrast, for the relational perspective, what we call 'emotions' are by-products of human interchange. Emotion terms (e.g., 'anger,' 'love,' 'depression') may serve as key elements of conversation (e.g., 'That makes me angry,' 'Do you love me?'). Yet, these terms are also embodied, in the sense that without certain patterns of facial expression, tone of voice, posture and so on, they would lose their intelligibility. In effect, we may say that emotions are forms of cultural performance (Averill, 1982). One doesn't possess an emotion so much as one engages in the doing of an emotion. The question is not, then, whether one is truly feeling love, sadness, or depression, but whether the individual is fully engaged in such performances.

At the same time, these embodied performances of emotion are also embedded within patterns of interchange. They acquire their meaning from their use within the ongoing process of relationship (Shields, 2002). I use the term 'relational scenario' in referring to the culturally sedimented patterns of interchange within which emotional performances may often play an important role (Gergen, in press). Thus, for example, the performance of *anger* (complete with discourse, facial expressions, postural configurations) is typically embedded within a scenario in which a preceding *affront* may be required to legitimate its meaning as anger. (One cannot simply shout out in anger for no reason; to do so would be to exit the corridors of intelligibility). Further, one's performance of anger also sets the stage for the subsequent performance of an *apology* or a *defense* on the part of another; and if an apology is offered, a common response in Western culture is *forgiveness*. At that juncture the scenario may be terminated. All the actions making up the sequence, from affront to forgiveness, require each other to achieve legitimacy. To function as a normal human being is to participate successfully within scenarios of relationship.

And so it is in the case of actions that we take to be voluntary. A person's claims to have chosen, intended, or tried are circumscribed by convention, not only in terms of the content of such avowals, the tone of voice, posture, gaze, and the like, but also in terms of the contexts in which they can be offered. And, once such avowals are offered, others are not free to respond in any way they wish. At the same time, such responses may also ratify or deny the avowal. To accept an apology because the 'other didn't mean to' ratifies the intention that is claimed by the other. To respond to an apology with 'I don't trust you for one minute; you knew very well what you were doing' denies the legitimacy of the apology. Essentially, the intention is not granted reality in the conversation. In the same way, an action on the part of one person cannot cause another's response, because the act in itself has no meaning. One can claim that 'my friends made me do it,' thus implying a cause and effect relationship. However, others may not ratify the causal ascription ('you could have chosen otherwise'), in which case the causal account is not sustained. Again, agency and causality are creations of coordinated action. Or, in Judith Butler's terms, 'there need not be a "doer behind the deed,"'... the "doer" is variably constructed in and through the deed' (1990, p. 142).

## Toward relational responsibility

These attempts to generate a relational conception of human nature are yet in their infancy. At the same time, their potentials are substantial. This is so both intellectually and in terms of socio/political reverberations. From the relational perspective, we confront the possibility of developing intelligibilities that go beyond the naturalization of separable units, animated by some interior *arche* or causal impingements. We understand distinctions between *me* versus *you*, and *we* versus *them* as contingent and potentially problematic characterizations of the human condition. Our concern shifts instead to the relational processes by which the very idea of individual units (selves, groups, and so on) comes into being. The focus moves from the dancers to the dance. And we are invited to consider alternative constructions and scenarios that may better serve human kind.

This brings us again to a central theme of this chapter: the case of human responsibility. What does a relational ontology offer as an alternative to traditional conceptions and practices of individual responsibility? To appreciate the possibilities we must first inquire into the origins of good and evil, and to do so in a way that neither invokes the vision of the individual evildoer, nor the hapless victim of heredity and environment.

As we have seen, the relational account traces the birth of meaning to coordinated action. The fruits of coordination also include orientations to value. Of special concern, as people become increasingly coordinated in their patterns of action, these patterns become accepted conventions. To unsettle the conventions is to threaten the traditions of coordination and all they bring forth. If such coordinations were perfected, there would be no untoward or problematic behavior. To violate the tradition – in effect, to do evil – would simply be unimaginable. In this sense, the primary glue that holds communities together is neither conscious deliberation of right and wrong, nor reward and punishment, but the sheer unintelligibility of doing wrong. Let us call this form of morality, inherent in virtually all coordinated relationships, first-order morality.

The creation of what we call evil is brought about through multiplicities of relationship. People typically participate in multiple relational traditions, thus sharing in the intelligibility of differing and often competing values. It is valuable to get ahead, but also promote equality, to care for the self but to take care of others, to be considerate and sympathetic but defend what is right, and so on. Evil action is not, then, the result of a warped conscience or lack of proper training, but is rendered 'a good thing to do' in at least some tradition. For example, robbery is a means of getting ahead, providing for one's family, and caring for the self. Murder is often the result of defending one's rights, territory, or preferred way of life. More critical cases of difference are those engendered by groups that share little in the way of values. Here it is entirely intelligible and righteous, for example, to drop atomic bombs on helpless Japanese civilians and to send planes crashing into the Twin Towers.

As I have argued previously, to hold individuals responsible in such cases invites alienation, antipathy, and retaliation. The challenge, then, is to locate an alternative to the presumption of individual responsibility, one more sensitive to the relational grounds of good and evil. Specifically, let us consider the potentials of *relational responsibility*, an orientation that places relational coordination at the center of concern (McNamee and Gergen, 1998). As might be ventured: all that we take to be good, valuable or moral issues from relationship. Thus, we may all be justifiably concerned with sustaining relational processes from which value is generated. The existence of evil is brought about by multiplicities of diverging achievements of the good. Thus, if we are to be responsible to the generative process of creating value, the tradition of holding individuals responsible is problematic. It primarily increases the divergence and thus the potential for mutual silencing. In effect, we move toward the end of meaningful action. The challenge of relational responsibility is thus to restore the impetus toward convergence, along with the potential for co-creating the good. In effect, the problems of evil created by a multiplicity of first-order moralities, invite the development of a second-order morality, one that enables alienated parties to engage together in the generation of first-order morality.

## Toward relationally responsible practices

There are numerous practices extant for the realization of individual responsibility, ranging from the informal ways in which we blame and punish, to judicial and prison systems, terrorism and warfare. So powerful is the tendency to hold the individual unit – person or group – responsible that retaliation now seems to approximate a genetic necessity. What forms of practice might issue, then, from a commitment to relational responsibility and the achievement of second-order morality? And could they yield more promising outcomes than the existing institutions of individual responsibility?

I can scarcely offer a practical panacea to the inevitable production of good and evil, nor a fully developed set of alternatives to existing traditions. However, emerging on a broad number of fronts are practices that place relational processes in the forefront of concern. There are two domains of practice that seem especially promising, the one emphasizing the interweaving of relationships, and the second, the dialogic softening of boundaries. In the following I shall outline the rationale for each, and touch briefly on several illustrative practices.

### Relational embedding as conflict prevention

The chief concern in this case is that of preventing the kind of solidification of relationship that renders others' realities and values irrelevant, and thus open to violation. Every relationship will tend to suppress or silence the voice of others. In some degree such silencing is necessary for coordinating any successful relationship. However, to the extent that other relationships are active participants in the relationship at hand, such suppression may be minimized. In Levinas's (1998) terms, we might say that when the voice of the other is present within us we are called into a condition of care. However, in more relational terms the practical problem is that of sustaining salience or the broader relational network in which one is embedded. When one's connectivity is real and apparent, it is more difficult to do harm to the other.

In surveying the landscape of emerging practices useful for vitalizing the relational matrix, the following are particularly promising.

*Organizational participation.* In organizational management practices, there is a continuing movement away from the kind of command and control practices of recent decades, to explorations of more participatory decision-making processes. Much discussion centers around the concepts of the flat organization and organizational democracy. However, one of the most effective practices to emerge from these concerns is that of *appreciative inquiry* (Cooperrider, Whitney and Stavros, 2003).

The practice of appreciative inquiry attempts to enlist the entire organization (insofar as possible) in charting its future direction. The process begins with pairs of organizational participants sharing stories about times they were most fulfilled, inspired, or nourished by their activities in the organization. Then, larger groups meet to share stories and to extract major themes or values. These results are then shared with a larger group to consider what kinds of practices or policies might be put in place to maximize these values. Committees are assigned specific roles to bring about the realization of these values. This practice has met with such broad enthusiasm that it is now practiced world wide. From the standpoint of relational responsibility, participants no longer see themselves as separate from the organization and their fellow participants, but as inseparable from the organization itself.

*Classroom participation.* In the educational sphere there is now a strong movement in the United States towards collaborative practices. These include a movement from monologic or top-down teaching to pedagogical practices of class-room dialogue (Barbules, 1993; Wells, 1999). Further attempts are being made to draw parents into deliberation with school officials about curriculum and classroom planning (Rogoff, Turkanis, and Bartlett, 2001). Most impressive is the movement toward collaborative writing, in which students work together on writing projects as opposed to separately (see, for example, www.stanford.edu/group/collaborate/). The broad result of such efforts is the creation of relational flows in which participants come increasingly to see themselves as embedded within the same efforts. It is not students against each other or the school, or the school as opposed to parents, and the like, but 'all of us participating together'.

*Therapeutic collaboration.* There has been growing criticism of the medical model, in which the therapist serves as 'the knowing one,' and treats the 'docile body' to bring about a cure. In effect, this traditional view of therapy is allied with the command and control approach to management, and the monologic form of education. In place of this individualist orientation is a view of therapy as a collaborative process. The therapist and client work together to create new realities. Most closely identified with this orientation are practices of narrative therapy (Angus and McLeod, 2004), brief therapy (De Shazer, 1994), and postmodern therapy (Anderson, 1997). However, a dramatic demonstration of relational practice is furnished by Seikkula and Arnkil (2006). In this case, the traditional process of diagnosis, in which the expert psychiatrist determines the 'nature of the illness' and the means to 'cure,' a dialogue takes place. This dialogue typically includes a social worker, the family of the individual designated as the patient, close friends of the 'patient,' and a range of others close to the problem (e.g., school officials, clergy). Through the ensuing dialogue, decisions are reached as to how best to proceed, and who will be responsible for the process. This practice has succeeded in lowering the number of people confined to mental hospitals and the prescriptions for medications.

In each of these domains we find, then, practical means of building relational interdependence. Such practices diminish the reality of the individual unit, the agent who is free to do as he or she wills, and create the alternative and potentially more compelling reality of 'we'. At the same time, by multiplying the range of these practices, the individual becomes a participant in multiple relations, at once more flexible and simultaneously capable of appreciating the many repercussions of his or her actions.

### Transformative dialogue as conflict dissolving

It is one thing to develop practices that lend themselves to a relationally responsible society. However, it would be utopian to think that society would thus be rid of activities deemed immoral, unethical, or abominable. And perhaps it is unrealistic to think that one day there might be little need for police forces, systems of individual justice, and prison penalties. We cannot simply leap over the shadow of history. However, in the case of wrongdoing there are now alternatives emerging to the traditional practices of individual blame. Rather than adjudicating blame and administering justice from a given standpoint, such practices are dialogic. Rather than the 'good people' versus 'the bad,' the attempt is to bridge the differences through alterations in both the process and product of conjoint meaning making. In effect, these practices of transformative dialogue are more responsible to the relational matrix from which the very possibility of moral action is achieved (Gergen, Gergen and Barrett, 2005).

To illustrate transformative dialogue at work, I will touch on three disparate practices.

*The public conversations project.*   Developed by a group of family therapists concerned with irremediable conflict (Becker et al., 1995), the project attempted to establish practices of communication that would enable otherwise alienated groups to live more peacefully together. The resulting practice includes first bringing members of groups together for a meal. Then, rules are established for an ensuing dialogue. Arguing about the principles behind one's movement (for example, pro-life vs. pro-choice) is forbidden. Instead, participants are invited to tell stories that bear on their commitment, to describe what is at the heart of the matter for them, and to describe any areas of doubt in their position. These dialogic practices prove highly successful in bringing about forms of understanding that enable participants to live together amicably, even recognizing their differences.

*Narrative mediation.*   Traditional mediation presumes a fixed reality with parties to the conflict functioning to maximize their own ends. With narrative mediation (Winslade and Monk, 2001) this individualist view is abandoned in favor of an approach that presumes socially constructed realities and unfixed ends. Through particular forms of questioning, the mediator enables the participants to see their conflict in terms of constructed stories, and to search together for new and more viable ways of understanding themselves and their relationship. Blame and retaliation are replaced by a joint search for a new and more connecting reality.

*Restorative justice.*   There is currently a globe spanning movement to replace traditional practices of blame and punishment for violent actions with collaborative interchange. The restorative justice movement (Umbreit et al., 2003; Miers and Willemsense, 2004) has been particularly concerned with cases of violence, and means of restoring the fabric of community as opposed to leaving an interminable rift of blame, resentment, and retaliation. Practitioners in this case typically work with both the victim and the offender, but with an interest in including all stakeholders in the matter. Through practices of mediation, the further attempt is to help the offender and the victim understand what has taken place from the other's point of view. Further, assistance is provided to offenders so that they may repair the harm they have caused, and for other stakeholders to enter into the process of resolving the injustice.

These few practices are only representative of a large body of inventive means of fostering and protecting the relational process. They are only a beginning, however, to a full flourishing of relational responsibility. One may raise the question, of course, as to whether these practices are ultimately attempts to control people's actions. Certainly Rose's contribution to the present volume (see Part II 'Self and (Socio-)Scientific Knowledge', Chapter 3) alerts us to the ways in which various social programs ('for our benefit') function to circumscribe and direct human action. However, issues of control are closely related to issues of first-order morality. As people come into coordination, and establish the value of their tradition, so will control become an issue. It is not an issue for those wholly within the tradition; they favor the form of life which they have fostered. It becomes a problem, however, for those who live – either partially or wholly – within another tradition. For them the commonly valued expectations and practices are endured or resisted. To resist all traditions of coordinated action, an invitation many find in Foucault's (1978, 1979) writings, would be to step outside the possibility of meaning altogether. And, such an attempt could not be undertaken except by following or 'capitulating to' the realities constructed within some tradition of meaning. In any case, the practices outlined here are essentially aimed at restoring the possibilities for the conjoint creation of value.

## In conclusion

The attempt of this chapter has been to press beyond the discourses of individual will and social causation, to an ontology of relational process. Such an ontology is especially invited by the problematic outcomes of holding individuals responsible for their actions. From a relational perspective any attempt by people to co-create a reality will be accompanied by a world of value. In this sense, the production of value is always already under way. The outcome of this process may be termed a *first-order morality*. As we have seen, however, a multiplicity of first-order moralities brings forth the existence of what we take to be immorality. When individuals (whether agentive selves or larger groups) are held responsible for their evil, it is in the service of restoring or affirming first-order morality. The strong risk, however, is that such practices of restoration do not diminish so much as stimulate the further production of evil. Required, then, is a *second-order morality*, one required by the potential ravages of first-order moralities. This second-order morality is essentially one of relational responsibility. As I have attempted to demonstrate, relational responsibility is most effectively invited through new forms of participatory practice.

# References

Anderson, H., *Conversation, Language and Possibilities* (New York: Basic Books, 1997).

Angus, L. E. and J. McLeod (eds), *The Handbook of Narrative and Psychotherapy* (Thousand Oaks, CA: Sage, 2004).

Aristotle, *Psychology*, trans. P. Wheelwright (New York: Odyssey Press, 1951).

Aristotle (350 BC), *Nichomachean ethics*, available at: http://classics.mit.edu/Aristotle/nicomachaen.html

Averill, J., *Anger and Aggression: An Essay on Emotion* (New York: Springer-Verlag, 1982).

Barbules, N. C., *Dialogue in Teaching* (New York: Teacher's College Press, 1993).

Barresi, J. and C. Moore, 'Intentional Relations and Social Understanding', *Behavioural and Brain Sciences*, IXX (1996): 107–45.

Becker, C., L. Chasin, R. Chasin, M. Herzig and S. Roth, 'From Stuck Debate to New Conversation on Controversial Issues: A Report from the Public Conversations Project', *Journal of Feminist Family Therapy*, VII (1995): 141–63.

Billig, M., *Arguing and Thinking*, 2nd edn (Cambridge: Cambridge University Press, 1996).

Butler, J., *Gender Trouble, Feminism and the Subversion of Identity* (New York: Routledge, 1990).

Cooperrider, D., D. Whitney and J. Stavros, *The Appreciative Inquiry Handbook: For Leaders of Change* (Cleveland, OH: Lakeshore Publishers, 2003).

Davidson, D., 'Freedom to act', in T. Hondrich (ed.), *Essays on Freedom of Action* (London: Routledge, 1973).

De Shazer, S., *Words Were Originally Magic* (New York: W. W. Norton, 1994).

Foucault, M., *The History of Sexuality Volume I: An Introduction* (New York: Pantheon, 1978).

Foucault, M., *Discipline and Punish: The Birth of the Prison* (New York: Random House, 1979).

Gergen, K. J., *Realities and Relationships* (Cambridge, MA: Harvard University Press, 1994).

Gergen, K. J., *Relational Being* (New York: Oxford University Press, in press).

Gergen, K. J., M. Gergen and F. Barrett, 'Dialogue: Life and Death of the Organization', D. Grant et al. (eds), *The Handbook of Organizational Discourse* (Thousand Oaks, CA: Sage, 2005), pp. 39–59.

Gibson, E. J., 'Are We Automata', in P. Rochat (ed.), *The Self in Infancy* (Amsterdam: Elsevier, 1995).

Levinas. E., *Otherwise than Being: Or Beyond Essence* (Pittsburgh: Duquesne University Press, 1998).

McNamee, S. and K. J. Gergen, *Relational Responsibility* (Newbury Park, CA: Sage, 1998).

Mele, A., *Autonomous Agents* (Oxford: Oxford University Press, 1995).

Middleton, D. and S. D. Brown, *The Social Psychology of Experience: Studies in Remembering and Forgetting* (London: Sage, 2005).

Miers, D. and J. Willemsense (eds), *Mapping Restorative Justice: Developments in 25 European Countries* (Leuven: European Forum for Restorative Justice Publication, 2004).

Pottor, J. A. and M. Wetherell, *Discourse and Social Psychology: Beyond Attitudes and Behaviour* (London: Sage, 1987).

Rogoff, B., C. G. Turkanis and L. Bartlett (eds), *Learning Together: Children and Adults in a School Community* (New York: Oxford University Press, 2001).

Rose, N., *Governing the Soul* (London: Routledge, 1990).

Seikkula, J. and R. E. Arnkil, *Dialogical Meetings in Social Networks* (London: Karnac, 2006).

Sheilds, S. A., *Speaking from the Heart: Gender and the Social Meaning of Emotion* (New York: Cambridge University Press, 2002).

Skinner, B. F., *Beyond Freedom and Dignity*, revised edn (New York: Hackett, 2002).

Umbreit, M. S., B. Bos, R. B. Coates and K. A. Brown, *Facing Violence: The Path of Restorative Justice and Dialogue* (Monsey, NY: Criminal Justice Press, 2003).

Wells, G., *Dialogic Inquiry, Towards a Sociocultural Practice and Theory of Education* (Cambridge: Cambridge University Press, 1999).

Winslade, J. and G. Monk, *Narrative Mediation: A New Approach to Conflict Resolution* (San Francisco: Jossey Bass, 2001).

Wittgenstein, L., *Philosophical Investigations*, trans. G. Anscombe (New York: Macmillan, 1953).

Wolf, S., *Freedom within Reason* (Oxford: Oxford University Press, 1990).

# 9

# The Role of the Self-Model for Self-Determination

*Tillmann Vierkant*

## Introduction: from free will to self-determination

In our modern world the language of freedom is everywhere. We live in a free country, we have the freedom to choose, the military forces of many Western states are fighting in an operation called 'Enduring Freedom' and in the United States, one could, at one time at least, even order freedom fries. Freedom is everywhere. Democracy is freedom: we are free to choose our government and if they do not do what we want them to do, then we are free to vote them out of office. The free market defeated its communist rivals, and it seems only a matter of time until we live in a world in which no trade restrictions will hamper our freedom to exchange whatever we want with whomever we want, as long as we freely decide to do so.

But while the language of freedom becomes more and more part of our daily lives, there remain at the same time nagging doubts about the existence of this freedom. In this context I am not referring to the many challenges to politicians and the market, to whether they can really deliver on the promise of freedom or whether they actually just bring new chains that make it just as difficult for us to live a free life as it always has been. These nagging doubts go far deeper than this, and there is no way that we can put the blame on the state or society or the capitalists. It seems that we might have more and more evidence that the whole rhetoric of freedom is empty, not because of others who take our freedom, but because of ourselves. The ancient doubt, supported by more and more empirical evidence, is simply that we ourselves cannot help to build a free society, because we are not free creatures.

It is not that anybody ever took our freedom, but that we never had it in the first place. This doubt is rooted in a theory that we might be nothing more than the aggregate of our genes and social conditioning: that is to say, there is nothing in us that would qualify us as real autonomous beings. Instead, all that exists are endless causal chains. Thus it seems as if all our decisions, from the most important to the most profane, have been made long before we even existed, when the world was set up to run its course. Such nagging doubts are extremely worrying, because they seem to take away one of the most essential conditions of all these important freedoms in our society, namely the capacity of individuals to freely decide whether they want to do the one or the other thing. Yet it seems that the doctrine of physicalist determinism renders such decisions impossible.

Without the ability to make such decisions, it seems as if the whole language of democracy and free markets doesn't make any sense. It seems rather pointless to have a freely elected leader, if she is elected by individuals whose decisions are not free anyway.

In philosophy there are two main solutions to this problem. Libertarians believe that the assumption of physicalist determinism is mistaken. Genes and education are important in determining who we are, but when it comes down to it, the individual always does have the freedom to decide whether it will or will not perform an action. Compatibilists, on the other hand, believe that physical determinism is compatible with, perhaps even necessary for, the idea of free will. By now, the debate between the two positions has produced enough material to fill several libraries and it does not seem as if the problem will be resolved any time soon. Therefore, I am not going to add another drop to this particular ocean by trying to offer any new material to the debate in this chapter. Instead, I am going to remold the question of the nature of the will into one that seems slightly less puzzling, and hope that the insights from this remolding might prove useful for obtaining a better grasp of what is at stake in the original problem.

Free will is a topic that most people find puzzling, but as we have seen above, this does not mean that they have difficulties at all in their everyday lives in using terms like 'freedom' and 'responsibility' quite consistently, even though they would be forced to admit, if pressed, that they have no idea how free will might be possible. Compatibilists see this as an argument in favor of free will actually being fully compatible with physical determinism, as otherwise we would not have this consistent practice. Incompatibilists, on the other hand, argue that the practice shows the falsity of physicalism: but both would have to agree that this practice is central for an understanding of what we mean by free will. The compatibilist and the anti-physicalist incompatibilist hold, for very different reasons, that this practice shows us free will in action, so to speak.

In everyday life, we describe an action as free if the actor wanted that action *herself*. One is free as a bird, because birds, being able to fly, can go wherever *they* want to go. We talk about our own free will, because the freedom of the will depends crucially on the fact that it is *us* who decide what we want to do and nobody else. In short, then, in practice free will can be translated into the idea of autonomous action or self-determined action. This is a neat enough remolding of the problem of free will, but it has one major disadvantage: it seems to substitute one problematic and mystical term with another, because who is the mysterious subject that determines itself? It is this question that is going to concern me for the rest of this chapter.

I will first introduce an idea of how one could make sense of the notion of self in modern naturalistic philosophy of mind and I will then apply this notion to the question of what we could mean, when we talk of self-determination, given the constraints of modern cognitive science. I will finally discuss one important objection to the account of self-determination that I sketch.

## Subject and self-model

The term 'self' seems every bit as ambiguous as the term free will. In every-day language, quite often our selves seem to be identical to the whole person, which here I take to be equivalent to the whole organism. When we say that we want to go to Japan, then we are not saying: My self wants to go to Japan, but I have to ask the rest of me first, whether she agrees. Utterances like this sound absurd. We are nothing else than the people we are. It seems nonsensical to divide the person into parts.

But our normal language does not always work like that: Sometimes we make a point of saying that our self is different from the body. For example, take all the stories about out of body experiences, in which people tell how they looked down at their bodies from the ceiling of the room. As well, there are stories about near-death experiences, where people talk about leaving their bodies and flying through a tunnel towards a distant light. Or there are the stories of people who have witnessed death, who described the experience as one in which they could see the soul leaving the body, and so on.

But we do not have only this kind of anecdotal evidence, which one might tend to dismiss as superstitious rubbish. Many eminent philosophers have joined ranks with our folk experience, agreeing that it seems that the human conscious mind is fundamentally different from its body. While the body has spatial extensions and behaves very much like other physical objects, the conscious mind seems to be fundamentally different – it is not spatial and it is not composed out of many small parts.

Additionally, it seems that the conscious mind is given to us in a way which is radically different from the way the rest of the world, including our bodies, is given to us. We can know things about the world and about our bodies, but we can get them wrong as well. Our minds, on the other hand, seem to be given to us directly and – to some degree at least – infallibly. It just does not seem to make sense to say: 'I experience pain, but I am not sure whether it is really me who is having this experience.' Once we get into this mode of talking, the gap between us as selves and our bodies seems to be very deep indeed.

In contemporary naturalistic philosophy of mind, most authors are very skeptical about whether there really might be a soul which could survive the death of the body. The general spirit of the time is materialistic and the arguments against dualism seem to be overwhelming. But even naturalistic philosophers have to account for the fact that in our phenomenal experience, it does not feel as if we are identical with our bodies.

In his 1993 book, *Subjekt und Selbstmodell* ( 1993), Thomas Metzinger develops a philosophical conception of subjectivity that is centrally based on the distinction between intentional systems (subjects) and their representations of themselves (self-models). In this book, Metzinger explains why it is that we believe that we are different from our bodies.

According to Metzinger, evolution developed self-models during a cognitive arms race, because they permit a tremendous number of new cognitive operations, for instance, long-term planning. For example, if an agent wants to do something that will benefit her, but not in the immediate future (e.g., carry some food home for dinner rather then going for a nice little nap), then that agent needs a sense of self in order to comprehend that she is doing the carrying for herself.

Nevertheless, it was not necessary in evolutionary terms that the system (subject) recognizes the difference between self-model and subject. This, according to Metzinger, results in the Cartesian illusion. We believe that we are our own representations of ourselves. Because of this belief, it seems to us as if we are Cartesian conscious minds. Whenever the subject (in Metzinger's terminology) feels, decides, or does something, it attributes this to its self-model. This self-model is nothing more than a specific functional role in the cognitive architecture of an information-processing subject, but this is not how it appears to the subject. For the subject, the nature of the self-model is invisible. The subject does not know that it is using a representation of itself in order to complete some high-level cognitive tasks. Instead, it seems to the subject that it is its own representation. It attributes all the work that is done by the enormously complex cognitive machinery of the subject to the in comparison, very primitive structure that is the self-model. Because of this error, it seems to us that our bodies are only tools that our immaterial mind uses to get things done, when really our seemingly immaterial minds are nothing more than a passive representational structure, highly useful for some tasks, but with absolutely no agency of its own. Instead, all agentivity lies with the embodied subject. The details of Metzinger's proposal are quite contentious, but do not matter here. All that is relevant for us is the argument that there is no immaterial agent over or underlying the acting body. So, given that the naturalistic equivalent of the Cartesian self, the self-model, does not seem to have any agentive qualities at all, it seems that the only suitable candidate for self determination has to be the subject in Metzinger's sense.

## The self-determination of the subject

In this section, I will try to answer the question of what it is about a piece of material (a body) that makes it an agent, without referring to immaterial selves and conscious minds. To find the answer, we have to go quite a long way down the evolutionary ladder.

Human beings, like all other living things, have the capacity to act in a goal-orientated manner. To some researchers, this ability is what makes organisms special. It is what makes them selves. Eleanor Gibson writes, for example:

> Living creatures have the power to control their movements and actions, and no planet or meteor or force of the physical universe does. As humans we have self-control, or agency, a far more remarkable kind of power than blind force. This is my candidate for how to think about the self – not a structure, or image of a body or a face, but control of one's actions and interactions with the world and with others.
>
> (Gibson, 1995, p. 5)

That the control of one's movement might be what gives organisms a sense of self is confirmed by quite a lot of empirical research. Motile organisms have to differentiate between their own movements and other movements, and if such organisms do not do this successfully, then they will suffer from illusions, because they will confound their own movements with those of other entities. An example of this is the illusion that the world changes when the eye is moved passively. The cognitive system is fooled, because it receives different inputs after the passive movement and cannot account for it in terms of its own movement. Therefore it 'concludes' that the world must have changed. Mechanisms to achieve this differentiation can already be found in very simple organisms like the Drosophila (Barresi and Moore, 1996). In higher animals, these feedbacks are more sophisticated and exist in various forms of proprioceptive and kinesthetic feedback.

The fascinating thing about these very basic mechanisms is that they explain something that philosophers had always taken as one argument for the fundamental difference between conscious mind and body: the intuition of privileged access, that is, the idea that our own agentivity is given to us in a fundamentally different way than the agentivity of others. The empirical evidence just cited shows that because of these monitoring mechanisms, one perceives action differently when it is one's own action than when it is the action of someone else.

The moral of the above seems to be that autonomy really is something quite basic once we look at it from the perspective of a cognitive scientist. Self-determination is what makes organisms agents that can pursue goals. Nevertheless, I will probably not have convinced most readers that I have successfully explained what the problem of human freedom really is about, and that actually it is not at all as complex as we always thought it was. This is not to say that what I have said so far has nothing to do with self-determination: being an agent in the sense described above is crucial to being an autonomous agent, but it does not seem sufficient. So what is it that is lacking?

### Awareness and self-determination

The following thought experiment from David Velleman serves very well as a tool to test our intuitions on what is needed for an action to count as willed. Imagine the following: you are disenchanted with an old friend of yours who recently started to make comments that seem cynical and abhorrent to you. You had decided to end the friendship for quite a while prior to meeting him again.[1] On this particular occasion, however, ending the friendship is not on your mind; you are simply glad to see him again. But while you talk to each other, you get more and more heated in the discussion and finally insult each other so badly that it leads to the breakdown of the friendship. Afterwards, you realize in a calm moment that your prior thoughts about your friend had given your arguments such a nasty and unfair edge, without you ever being aware of it, that your friend had little choice but to end the friendship. You feel you missed a great chance to mend fences with your friend, and are very sad that your remarks led to the final breakdown of the friendship.

According to what we have said so far, this example should describe a prototype of a free action.

I have claimed that what is special about self-determination is the fact that organisms can control their actions by means of goals they have. Now, in the thought experiment above, you certainly had formed the goal to end the friendship, and this goal was certainly influential in bringing about the intended result, even though you were not aware of it at the time.[2] If this is true, though, then the described example should be a prototypical intentional action.

But certainly this cannot be the correct interpretation of the thought experiment. It is true that you had intended to end the friendship, and it is true as well that this made you nastier than you would have otherwise been, but you did not want to be nasty to end the friendship! This is why you describe the situation passively. It was your goal and not you that severed the friendship. In the conversation you did not want to do anything about the friendship, but that something happened nevertheless was certainly a consequence of your prior goals, but not ones that you were conscious of. You cannot claim that you had 'nothing' to do with the action, but it did not have your full approval either. The result of your meeting was an accident. Surely such a situation differs massively from one where you consciously make the remarks in order to provoke the breakdown of a friendship? While you can't escape full responsibility in the first case, because your action was after all caused by some goals that you had consciously entertained before, you are obviously not responsible for such an action in the same way as in the second case, where you are fully aware of what you are doing.[3]

The surprising consequence of all this is that the self-model which we previously rejected as the agent of our voluntary actions, because taking it as such would lead us to the Cartesian illusion, has now staged a come-back in a more roundabout way. It might be true enough that our self-model is not what does all the cognitive work in information processing systems like us, but it seems that its presence is nevertheless a necessary condition for autonomy.

If we think about it, this is not really surprising either. For autonomous behavior it is not enough to have goals; you have to know that you have them. How else would you be in a position to change them? In addition, if it is impossible to change them, how could you be responsible for them? This is where the self-model is crucial. It allows the agent to understand her goals as her own which in turn enables her to reflect on the question whether it is positive that she has the goals she has.

Nevertheless, apart from the fact that the self-model has sneakily made itself very important again, we do not have a problem, or so it seems. We now understand the relationship between subject and self-model and we understand that in line with our folk psychological intuitions, self-models are actually crucially important for autonomous actions because they allow the subject to reflect on her own goals.

## The self-model and compatibilist accounts of autonomy

The great advantage of reintroducing the self-model in the self-determination story is that it actually ties up quite nicely with the most influential compatibilist idea about free will. Harry Frankfurt (Frankfurt, 1971) introduced in his seminal article 'Freedom of the Will and the Concept of a Person' the idea that free actions are those actions that are in line with the picture that one has of oneself. Frankfurt gives the following example: A drug addict might be quite capable of acting in a rational, purposeful and very complex manner to get the drugs she needs for her addiction. Nevertheless, this person might not experience her actions as self-determined. Like in the Velleman case, she has the feeling that the goals that drive her behavior are not hers. She feels estranged from them. She does not want to have the wants that drive her behavior. Frankfurt argues that it is these wants about wants which determine whether an action is really free and that only if we want to have the wants that drive our behavior are we really free. These wants about wants are referred to as second-order volitions

By adding the Frankfurtian picture in this way, it seems we now have a picture of self-determination which is empirically plausible and in line with our folk psychology, as well as with influential philosophical concepts of self-determination and does justice to the importance of consciousness.

All I had to do to get this result was resurrect our common-sense picture of autonomy in a materialistic framework. There is a problem with this otherwise very satisfying picture, though. The bad news is that if this picture is correct, then there is some striking empirical evidence which seems to make it very unlikely that we are self-determined in the sense just described: The research of Benjamin Libet (1985), and following on from this, the advocates of the 'adaptive unconscious' (Wilson, 2002).

## Self-determination and the self-model

Probably the most famous challenge posed by the cognitive sciences to our folk psychological ideas about voluntary action comes from Benjamin Libet ( 1985). In his ingenious experiments, Libet tried to test the temporal relationship between the conscious awareness of a decision to act and the so-called 'readiness potential' (RP), an unconscious electrical potential which precedes every voluntary motor action. To accomplish this, he asked subjects to flex their wrist, whenever they felt like it, within a set period. At the same time, subjects were told to look at a very exact chronometer and asked to note the exact time when they became aware of their decision to flex. His results showed that this conscious awareness preceded the actual movement onset by about 150 milliseconds. More importantly, however, it turned out that the unconscious RP preceded conscious awareness by a staggering 300 milliseconds. Libet himself argued that his findings did not rule out a constructive role for conscious free will. The 150 milliseconds before the actual movement onset could be used to exercise a conscious veto (even though he admits that there is no proof other than introspection that the conscious veto might not itself be unconsciously initiated). According to Libet, then, movements are unconsciously initiated, but consciously monitored and vetoed.

Many critics were not quite satisfied with Libet's interpretation of his own experiment. They pointed out that the empirical evidence for the veto is very small and pretty dubious. But even if one granted the possibility of a veto, the scope of its influence would be very narrow indeed. Patrick Haggard and Martin Eimer (1998) demonstrated that the general story of the Libet experiments also works for choice paradigms. In their experiments, subjects could freely decide whether to flex the left or the right hand. In this case, the conscious awareness to flex left or right was preceded by a lateralized RP, indicating that the decision to flex left or right was made before it entered consciousness. If this holds true, then the conscious veto seems to have very limited powers indeed. It would mean that consciousness is not able to choose which action to perform, but only to veto the unconscious choices.

Even this form of veto is unlikely, however. On the one hand, findings by Hakwan Lau and colleagues (2004) seem to show that awareness might actually only occur *after* the movement has already begun. Lau argues that the timing in the Libet experiments is just down to a mistake in the psycho-physiological methodology.

More interesting still are the findings by Daniel Wegner (Wegner, 2002). In his research, he has been able to show that it is possible to manipulate the sensation of voluntary control in his subjects. In his set-up, Wegner managed to convince people that they had willed a particular action – even though they had in actual fact been forced to perform it – providing they had a suitable explanation for why they might have wanted to perform the action. According to Wegner, this shows that the conscious feeling of willing is not causally relevant in action execution, but rather the interpretation of some ongoing subpersonally-initiated behavior.

Does this evidence show that there is no autonomy? One could think so. At the very least, this evidence shows that there is a problem with the model developed in the last section, according to which the conscious self-model has to cause actions for them to be self-determined. If this model is correct (and the arguments for it seemed quite convincing at the time), and if all this evidence from the cognitive sciences really shows that the self-model is not involved in the causation of actions, then it follows that there are no self-determined actions at all.

But does it really? The problem here might be a wrong, or at least very limited understanding of causality. Fred Dretske's (1988) distinction between 'triggering' causes and 'structuring' causes can be used to explain what is at issue here. Here is an example of how to understand this distinction: if you want to water your garden with a hosepipe, you have to open the tap to trigger the water to flow, but opening the tap will not help, if the pipe is not structurally sound, that is, if the pipe has holes. More generally, for a specific event to happen, many causes have to be at work. Some of these causes directly trigger a specific event, others are more persistent They are structures set up to cause many events of a specific type, but they tend not to be involved in the specific timing of one particular event, for which the trigger is responsible.[4]

One way to escape from the dilemma posed by the Libet experiments is to resort to the idea that conscious self-models work more like structures than triggers. They are not there to initiate specific motor behavior, but they function as constraints for the unconscious processes that do. Once we allow for such an understanding, the threat the Libet experiments initially seemed to pose to our folk psychological notions of self-determination simply collapses. Even if it is true that the self-model is not a trigger for our actions, this does not mean that we cannot be autonomous, because the self-model can still causally influence the production of an action by acting as a relevant constraint.

So far we have seen why the conscious self-model is so crucial for self-determination, even if we understand the conscious self not as an immaterial agent, but just as a specific set of representations used by the embodied subject to tell itself a story about who she is and what is the right thing for her to do. We managed to overcome the challenge posed by the cognitive sciences that such a self does not participate in action generation and can therefore not be what makes us free – but this is not the last, nor even the most severe challenge which can be mounted against these conclusions. It is possible to argue that even though we established that conscious goals can be a tool for self-determination, it is not at all clear whether they should be.

## Are conscious reasons all there is to self-determination?

Consider the classic case of Huckleberry Finn. Huck thinks he knows of many reasons why it is morally wrong to help slaves on the run. He is not aware of any good reasons supporting the position that such an action might be justified. But when he finally gets the chance to act on his reasoning and to turn the escaped slave Jim in, he does not do it. He acts on an impulse and lets Jim go. This example is a prime case of a moral decision against what one believes to be the reasons that one has for acting.

The Huck Finn case shows that it is not obvious that consciously entertained reasons are really the only ones that count as true reasons of the actor. There is a strong intuition that Huck Finn discovered something about his true self when he allowed Jim to escape. On the conscious reason account, though, it seems as though we have to describe the case as one where Huck is overpowered by some strange compulsion. His act is no more free than the act of the unwilling drug addict.

The idea that Huck discovers his true self in this situation does actually make a lot of sense if we return to the naturalistic story that I told above about the relationship between subject and self-model. After all, the self-model is only a model of what is important for the subject, and like all models, it can be more or less adequate. Discovering that the self-model is faulty should surely be agency in the best sense of the word, and not just an example of compulsive behavior.

Do cases like Huck Finn show that the story about the importance of conscious reasons I have told so far is seriously flawed? I don't think that this is the case. To see why, we have to look at the example a little more carefully. Philosophical thought experiments are devilish things, and sometimes it is not quite clear what exactly triggers our intuitions.

The best antidote to a thought experiment is quite often another one. Consider, for example, the case of Hick Fun. Like Huck, Hick is traveling with the escaped slave Jim, but unlike Huck, Hick has read many books on universal human rights and believes that these books show convincingly that slavery is morally wrong. Nevertheless, when the time comes, Hick betrays Jim on a spontaneous impulse. He just acts on a spontaneous feeling that black people are too primitive to look after themselves and a danger to society if left unobserved. Hick is actually not a very untypical person in this. In an extremely disturbing piece of research, Tony Greenwald and his collaborators (1998) could show that about two-thirds of the population have unconscious stereotypes, according to which black people are more closely associated with weapons than white people. These stereotypes not only passively exist somewhere in the psyche of the people that took part in the research, but have been shown to be causally effective in the production of behavior. Particularly disturbing about this research is the fact that these stereotypes were not only recorded in people with strong conscious prejudices against black people. One of the groups that scored above average for unconscious stereotypes were NGO aid workers. Are we willing to claim that actions triggered by these stereotypes show something about the true self of these aid workers? It seems not. Rather, it seems to me, what we would like to say is that in cases like these, the NGO workers, like most of us, were more influenced by the enormous flood of pictures and stories about 'dangerous black people' than they thought possible. They might be aware of these stereotypes within themselves, or they might not be, but they were very probably convinced that they had them under control, because of their consciously acquired reasons about human equality.

### So why do we intuitively feel that Huck exercises his freedom?

So if in this case we are not likely to make the stereotypes part of the person and will understand them instead as sub-personal mental content which threatens self-determination, then we have to ask ourselves why we felt different in the structurally identical case of Huck Finn?

The evaluative asymmetry used in my case studies is actually nothing new. They are very similar to cases that have been used to illustrate what Susan Wolf calls the 'asymmetry thesis' (see, for example, Fischer and Ravizza, 1998). In many cases, we find it perfectly acceptable to praise somebody for an action, even if that person could not consciously control her behavior, but it seems wrong to criticize somebody for a specific behavior, if the person could not control her behavior.[5] Think, for example, of a woman who is psychologically determined to jump into water to help a drowning child. Surely, she is praiseworthy for her actions. That she was determined to act in this way, if anything, enhances her moral stature. Now think of a woman who is hydrophobic and therefore psychologically unable to save the child. Surely, she cannot be criticized for her behavior. She might have wanted to help, but her pathology made it impossible for her. If anything we should pity her for the horrible state she must be in, after watching a child drown without being able to help.

The asymmetry thesis seems to provide a perfect explanation for the asymmetry in my two cases.

The problem with this explanation is that the asymmetry thesis itself is quite contentious. John Fischer has doubted that the asymmetry thesis really is intuitively plausible. He argues that it only seems to work in the example mentioned above because there is a reason-responsive mechanism in the praise sequence, but a reason-undermining one in the 'bad' sequence (Fischer and Ravizza, 1998). That is to say: the psychological determination to do good seems of a different kind to a phobia. The person is determined, because she is able to translate her knowledge of what is the right thing to do into a powerful disposition to act on that knowledge. This link to the reasoning capacity is absent in the hydrophobia case, and that is what distinguishes the two cases.

This seems to me a valid argument against the case discussed above, but does the same argument work for Huck and Hick?

It seems not, because the two cases seem identical with respect to reason-responsiveness. Both cases are cases of spontaneous actions, which are not in accordance with the explicit moral reasons of the actor, but expressions of some unconscious cognitive structure. Whether this structure is reason-responsive or not depends on a more exact definition of reason-responsiveness. They are not reason-responsive insofar as the actor could not explain why he acted the way he did, but this does not mean that these structures might not change, if they were not backed up by additional external inputs from time to time.

Nevertheless, Fischer's argument is not toothless in this case either. Because even though Huck and Hick might be the same with respect to reason-responsiveness, both cases are ambiguous at the same time. Both can be understood as products of compulsive behavior that the agent would produce even if it would be very irrational for him to do so (e.g., if the behavior were followed by instant death), or as a 'clever,' flexible cognitive mechanism the agent is simply not aware of. Perhaps, Fischer could argue, what produces the seeming asymmetry is the fact that because Huck does the right thing, whereas Hick acts in the wrong way, we tacitly assume that Huck's mechanism was probably reason responsive, whereas Hick's was not.

One might illustrate this with the following version of the Huck story: If we found out that Huck only helps Jim because he has an unconscious pathological fear that every black man will kill people who betray him, but would have happily turned in any Asian or Indio slaves, then we would not feel that his spontaneous decision to act against his consciously available reasons was really autonomous. A pathology is not something that makes us free, even if it leads to the morally required action in a particular situation. Once we have made it clear that Huck's mechanism was not really reason-responsive, the intuition that he acted autonomously vanishes in an instant.

But this seemingly harmless satisfying solution of the Hick\Huck asymmetry is bad news for my account of autonomy. If it is correct, then we have now an argument against the explicitness condition for responsibility that I have advocated.

One could take the Fischer-style interpretation of the Huck story to show that reason-responsiveness is all that matters for autonomy, whether it is conscious or unconscious. The intuitive plausibility of the explicitness condition could be explained by the fact that explicit processing is associated with our capacity to reason.

This is an interesting challenge, but I think it fails. First of all, Fischer's explanation does nothing to explain the intuition in the Velleman example, which provided the reason in the first place for demanding the explicitness condition at all; but I think the Huck example is no counter-argument either. What the Huck/Hick example shows is that reason-responsiveness is a necessary condition, but it does not show that it is sufficient. That is, it is not clearly demonstrated by the Hick example, or more importantly by our intuitive reaction to learning about the existence and causal efficacy of unconscious stereotypes in all of us. These stereotypes might indeed be quite reason-responsive, but we certainly do not always feel that behavior caused by them is autonomous behavior.

But even if this should be a good enough reason to reject the challenge to the explicitness condition, we still haven't explained the asymmetry in our intuitions towards Huck and Hick.

## The normative dimension

There is an important ambiguity in the notion of reason-responsiveness, and I think this ambiguity explains the asymmetry in our intuitions. Reason-responsiveness might refer to a coherence of subjective reasons, or it could refer to an ideal coherence of the objectively best course of action in a particular situation.

If we understand reason responsiveness in the first sense, then there is no asymmetry between Huck and Hick, but this changes dramatically once we employ the second notion – for on this notion of reason-responsiveness, Hick's stereotypes are *not* reason-responsive, even though they might fulfill all the subjective rationality conditions, because we are convinced that these stereotypes are objectively wrong.

If we employ this second notion of reason-responsiveness, however, then self-determination is affected as well, because now there is nothing in the organism at all any longer which decides whether an action is autonomous or not. The structurally identical mechanisms in Huck and Hick are evaluated differently solely because they stand in a certain relationship to the objective world: Huck's mechanisms pick up a real feature of the world, whereas Hick's mechanisms err.

The self that determines here is not only different from the self-model, but is also even larger than the subject. This self is an ideal relationship between the organism and its natural and social environment. It is this notion of self that we might have in mind when we talk about finding our 'true' selves, or which is expressed in the classical phrase 'know thyself'.

Defining our understanding of the meaning of self-determination as the determination of behavior by this kind of self is very appealing. Here, it is truly what is in the best interest of the subject that determines how she acts. This self seems immune to all kinds of manipulation, such as hypnosis, well-meaning neurosurgeons and the like, which certainly pose problems for the self-model understanding of self-determination. The self-model account on its own does not seem to have sufficient resources to distinguish such manipulations from true self-determination. This is different for what I would call the 'true self' account: whenever these manipulations are present, they are either in line with the ideal relationship between organism and environment – in which case it seems strange to describe them as manipulations rather than as autonomy enhancers – or they are not, in which case their presence per definition undermines self-determination.

## Can the true self really determine anything?

There is, however, one obvious problem with this last account of self-determination. It is unclear how this self is supposed to do any determining at all. The true self does not seem to be in any way connected to the causation of behavior. This fact creates dire prospects for the very possibility of self-determination. If the true self cannot cause behavior, then it becomes a very counterintuitive claim that it could have a role in self-determination. So, even though the true self account would be the ideal candidate as a sufficient condition for ideal self-determination; as it stands, it is actually no help at all and we are left with our slightly less perfect, but workable idea of the self-model account.

Nevertheless, there might be an interesting way in which the true self could be understood as relevant for self-determination. In order to accomplish this, we have to change our definition of the true self again. Instead of defining the true self as an abstract ideal relationship, we could define it as a specific set of causal relationships between the organism and its environment.[6] One could, for example, argue that the true self is a set of causal determinants, which are established within a social context. Stereotypes a society deems useful (i.e., a stereotypical help response towards children) might constitute such building blocks of the true self. That stereotypes are able to cause behavior seems unproblematic, but how can they be understood as self-determination? This only makes sense if we allow for some kind of reasoning capacity that goes beyond the cognitive apparatus. If social groups, for example, could have some such capacity, then it might make sense to speak of behavior of an individual that is self-determined in the sense that it is caused by a mechanism, which the individual has because society finds it appropriate to have it.

This could certainly be an interesting move, but whether it can be made plausible remains to be seen. For present purposes, the true self can at least be ruled out as an autonomy enhancer for a self-determining subject.

## Summing up

This chapter was able to show that it is possible to make sense of self-determination, even if we do not believe in substance dualism. By differentiating the different self-types, we could see clearly what kind of self-determination could be a basis for autonomy under the conditions of modern-day naturalism. The differentiation helped to get rid of the Cartesian homunculi that arise when we misunderstand conscious self-representations as acting subjects. Nevertheless, I was able to demonstrate that self-models do play a central role in the autonomous actions of subjects and that we only rate actions as autonomous, if they are in line with our self-models. We could see as well that the exciting findings of the cognitive sciences do not make it impossible to construct a causal role for the self-model in the action-generation of the subject, although the role of the self-model must be more indirect then the classical Cartesian picture assumes. Self-models are not triggering causes of voluntary actions, but that does not mean that they cannot be structuring causes, which constrain the space available for possible actions to create faster triggering systems. I did not have room within the bounds of this chapter to discuss possible practical consequences of such a picture of self control, but this does not mean that there are none. There might be pressing reasons for the discourse of law to attach responsibility to singular actions, regardless (within certain limits) of the perpetrator's history; but, for our moral considerations, it seems worthwhile to ponder the thought that if in concrete situations it is actually a fast unconscious system that initiates behavior and not the slow conscious one, and if it is true at the same time that the deliberative system merely, as it were, configures the fast system on a longer time scale, then it seems justified to emphasize such configuration processes more for our idea of responsibility. Virtues, for example, seem to be a typical example of successful configuration processes, and should therefore be allowed to play a far bigger role in our moral thinking.

The second half of the chapter discussed an important challenge to the reductionist, internalist picture painted in the first half. Starting with the problem of Huck Finn, it could be shown that we do have intuitions that reach beyond the structural make-up of the cognitive machinery of an individual. Even though Huck and Hick have analogue cognitive structures, our intuitions seem to tell us that Huck acts autonomously, while Hick does not. The chapter argued that the explanation for this phenomenon should be looked for in a third form of self-determination, where the self is not reducible to self-model or subject. This self, which I termed the 'true self,' is a normative entity, which is in a sense radically independent of the knowledge or capacities of the individual.

Whether this true self can be understood to be relevant for some form of self-determination had to remain open. The only thing that could be shown is that taking the true self to be relevant has the radical consequence that self-determination is not self-determination because of a particular individual cognitive system. On such a view, one would have to allow for the possibility that two physically identical beings in different environments could differ with respect to their status as autonomous beings. Whether such an idea could be made plausible is certainly material for much further research. For this chapter, it seems adequate to close on the note that it is quite possible to give a naturalistic account of autonomous action, even though much work remains undone.

## Notes

1 Velleman's original example does make slightly stronger assumptions than my experiment here. Velleman believes that the thought example would be sufficient, even if you had never been aware of your intention to sever the friendship. I have opted for slightly weaker assumptions, because I think they are enough to prove my most important point – that the agent was not involved in the action sufficiently for it to count fully as her action.

2 It is not a case of the infamous deviant causal chains either. Deviant causal chains are about special cases of behaviour that the standard story admits to, but maintains are very rare. In such cases, an intention is causally influential in bringing about an action, but not in the way it is supposed to. Donald Davidson's ( 1973) example of a climber who becomes so nervous, because he knows that he would like to let go of the rope, that he does in fact let go is the standard example. The 'friendship' scenario is importantly different, however. In this scenario, it is not the sheer existence of an intention that triggers – in a very unusual roundabout way – involuntary behaviour, but the intention that influences the action very much in the normal way, i.e., in the way that intentions control our behaviour in normal cases of actions.

3  Here I depart from Velleman's interpretation of his example. According to Velleman, his example shows that an action is only an action if the desire to act rationally plays a role in the decision-making process, but it seems to me that this does not solve the puzzle. It seems quite possible that the actor in this scenario would feel that he did not actively and freely end the friendship, even if he had decided beforehand that it would be rational to do so, thereby resolving an internal motivation conflict between nostalgia about the good times and annoyance about the remarks the friend makes every time they meet. He might even be relieved that the friendship ended the way it did, because he might still feel that it is better, all things considered, but might well insist nevertheless that he did not want to do what he did, because at the time he did not want to act on his reasoning. What is missing in this rationality story is the conscious awareness of the intentions that were causing his behaviour.
4  Additionally there are the background conditions, like the law of gravity etc.
5  Wolf is actually talking about the ability to do otherwise. I have tried to translate her example into my control terminology in order to avoid getting entangled in the ability to do-otherwise debate, which is not my concern here.
6  See Hurley (2004) for interesting ideas on this point. Gergen (Chapter 8 in this volume) seems to be hinting at a similar line with his idea of relational accounts of responsibility.

# References

Dretske, F., *Explaining Behaviour: Reasons in a World of Causes* (Cambridge, MA: MIT Press, 1988).

Fischer, J. and M. Ravizza, *Responsibility and Control* (Cambridge: Cambridge University Press, 1998).

Frankfurt, H., 'Freedom of the Will and the Concept of a Person', *Journal of Philosophy*, LXVII, 1 (1971): 5–20.

Greenwald, A. G., et al., 'Measuring Individual Differences in Implicit Cognition: The Implicit Association Test', *Journal of Personality and Social Psychology*, LXXIV (1998): 1464–80.

Haggard, P. and M. Eimer, 'On the Relation Between Brain Potentials and Voluntary Action' (MPI Paper, 1998).

Hurley, S., 'Imitation, Media Violence, And Freedom of Speech', *Philosophical Studies*, CXVII (2004): 165–218.

Lau, H., et al., 'Willed Action and the Attentional Selection of Action', *Neuroimage*, XXI, 4 (2004): 1407–15.

Libet, B., 'Unconscious Cerebral Initiative and the Role of Conscious Will in Voluntary Action', *The Behavioral and Brain Sciences*, VIII (1985): 529–66.

Metzinger, T., *Subjekt und Selbstmodell: die Perspektivität phänomenalen Bewusstseins vor dem Hintergrund einer naturalistischen Theorie mentaler Repräsentation* (Paderborn: Schöningh, 1993).

Wegner, D. M., *The Illusion of Conscious Will* (Cambridge, MA: MIT Press, 2002).

Wilson, T., *Strangers to Ourselves: Discovering the Adaptive Unconscious* (Cambridge MA: Belknap Press, 2002).

# Index

227